Climatic and Environmental Significance of Wetlands:

Case Studies From Eurasia and North Africa

Abdelkrim Ben Salem
Faculty of Sciences, Mohammed V University, Rabat, Morocco

Laila Rhazi
Faculty of Sciences, Mohammed V University, Rabat, Morocco

Ahmed Karmaoui
Moulay Ismail University, Meknès, Morocco & Moroccan Center for Culture and Science, Morocco

A volume in the Practice, Progress, and Proficiency in Sustainability (PPPS) Book Series

IGI Global
PUBLISHER of TIMELY KNOWLEDGE

Published in the United States of America by
IGI Global
Engineering Science Reference (an imprint of IGI Global)
701 E. Chocolate Avenue
Hershey PA, USA 17033
Tel: 717-533-8845
Fax: 717-533-8661
E-mail: cust@igi-global.com
Web site: http://www.igi-global.com

Library of Congress Cataloging-in-Publication Data

Names: Ben Salem, Abdelkrim, 1981- editor.
Title: Climatic and environmental significance of wetlands : case studies from Eurasia and North Africa / Abdelkrim Ben Salem, Laila Rhazi, Ahmed Karmaoui, editors.
Description: Hershey, PA : Engineering Science Reference, [2022] | Includes bibliographical references and index. | Summary: "This book investigates global changes impacts on ecosystem services and biodiversity in wetlands and offers concepts, methodologies, tools, and applications for ecosystem services valuation, wetlands biodiversity conservation, fresh water supply and its impact on biodiversity, agricultural production, food security and wetland management"-- Provided by publisher.
Identifiers: LCCN 2021052905 (print) | LCCN 2021052906 (ebook) | ISBN 9781799892892 (hardcover) | ISBN 9781799892908 (paperback) | ISBN 9781799892915 (ebook)
Subjects: LCSH: Wetland biodiversity--Eurasia. | Wetland biodiversity--Africa, North. | Biodiversity--Climatic factors--Eurasia. | Biodiversity--Climatic factors--Africa, North. | Wetland ecology. | Wetland biodiversity conservation. | Wetland management.
Classification: LCC QH541.5.M3 C55 2022 (print) | LCC QH541.5.M3 (ebook) | DDC 577.68--dc23/eng/20211104
LC record available at https://lccn.loc.gov/2021052905
LC ebook record available at https://lccn.loc.gov/2021052906

This book is published in the IGI Global book series Practice, Progress, and Proficiency in Sustainability (PPPS) (ISSN: 2330-3271; eISSN: 2330-328X)

British Cataloguing in Publication Data
A Cataloguing in Publication record for this book is available from the British Library.

Practice, Progress, and Proficiency in Sustainability (PPPS) Book Series

ISSN:2330-3271
EISSN:2330-328X

Editor-in-Chief: Ayman Batisha International Sustainability Institute, Egypt

MISSION

In a world where traditional business practices are reconsidered and economic activity is performed in a global context, new areas of economic developments are recognized as the key enablers of wealth and income production. This knowledge of information technologies provides infrastructures, systems, and services towards sustainable development.

The **Practices, Progress, and Proficiency in Sustainability (PPPS) Book Series** focuses on the local and global challenges, business opportunities, and societal needs surrounding international collaboration and sustainable development of technology. This series brings together academics, researchers, entrepreneurs, policy makers and government officers aiming to contribute to the progress and proficiency in sustainability.

COVERAGE

- ICT and knowledge for development
- Eco-Innovation
- Strategic Management of IT
- Innovation Networks
- Green Technology
- Environmental informatics
- Global Business
- Sustainable Development
- Technological learning
- Knowledge clusters

IGI Global is currently accepting manuscripts for publication within this series. To submit a proposal for a volume in this series, please contact our Acquisition Editors at Acquisitions@igi-global.com or visit: http://www.igi-global.com/publish/.

The Practice, Progress, and Proficiency in Sustainability (PPPS) Book Series (ISSN 2330-3271) is published by IGI Global, 701 E. Chocolate Avenue, Hershey, PA 17033-1240, USA, www.igi-global.com. This series is composed of titles available for purchase individually; each title is edited to be contextually exclusive from any other title within the series. For pricing and ordering information please visit http://www.igi-global.com/book-series/practice-progress-proficiency-sustainability/73810. Postmaster: Send all address changes to above address.

Titles in this Series

For a list of additional titles in this series, please visit:
www.igi-global.com/book-series/practice-progress-proficiency-sustainability/73810

Socio-Economic Sustainability in the Post-Pandemic Era
Jozef Oleński (High School Technology and Economics, Jaroslaw, Poland) Jeffrey Sachs (Columbia University, USA) Masayuki Susai (Nagasaki University, Japan) Yannis Tsekouras (University of Macedonia, Greece) and Arjan Gjonça (London School of Economics and Political Science, UK)
Information Science Reference • © 2023 • 315pp • H/C (ISBN: 9781799897606) • US $215.00

Climate Change, World Consequences, and the Sustainable Development Goals for 2030
Ana Pego (Nova University of Lisbon, Portugal)
Engineering Science Reference • © 2023 • 306pp • H/C (ISBN: 9781668448298) • US $250.00

Positive and Constructive Contributions for Sustainable Development Goals
Cristina Raluca Gh. Popescu (University of Bucharest, Romania & The Bucharest University of Economic Studies, Romania)
Engineering Science Reference • © 2023 • 281pp • H/C (ISBN: 9781668474990) • US $245.00

Leadership Approaches to the Science of Water and Sustainability
Kristin Joyce Tardif (University of Arkansas, Fort Smith, USA)
Engineering Science Reference • © 2022 • 219pp • H/C (ISBN: 9781799896913) • US $215.00

Eco-Friendly and Agile Energy Strategies and Policy Development
Mir Sayed Shah Danish (University of the Ryukyus, Japan) and Tomonobu Senjyu (University of the Ryukyus, Japan)
Engineering Science Reference • © 2022 • 272pp • H/C (ISBN: 9781799895022) • US $240.00

For an entire list of titles in this series, please visit:
www.igi-global.com/book-series/practice-progress-proficiency-sustainability/73810

IGI Global
PUBLISHER of TIMELY KNOWLEDGE

701 East Chocolate Avenue, Hershey, PA 17033, USA
Tel: 717-533-8845 x100 • Fax: 717-533-8661
E-Mail: cust@igi-global.com • www.igi-global.com

Table of Contents

Detailed Table of Contents

Wetlands around the world face anthropogenic and climate pressures that threaten key ecosystem processes such as the water cycle. Such threats are most acute for non-perennial systems that rely on occasional surface water flows to maintain important ecological functions. This work will focus on the use of remote sensing within a cloud-based computing platform to assess the spatio-temporal evolution of Dayet Aoua through the development of a water transition map between 1984 and 2019. It is also question of detecting and spatializing the main drivers of the various challenges, mainly the land use changes operated. This study revealed that a combination of changes in climatic and land use change in the vicinity of the lake has recently affected the dynamic and the recovery capacity of Dayet Aoua. This study highlights the need to increase public awareness and decision-makers' knowledge concerning the relationship between land use change and wetlands degradation.

Monitoring land use / land cover changes is essential for planning and management activities to conserve a particular habitat. In this study, the authors mapped land use / land cover changes in the Gediz Delta (Turkey) between 1984-2020 using Earth Observation and Geographic Information Systems. The maps were built upon the Horizon-2020 satellite-based wetlands observation service processing methodology and algorithms. The authors compared changes inside and outside the Ramsar Area in the Gediz Delta. The results indicate more than 147% increase in built-up areas and decreases of 33% in natural wetland habitats and 27% in natural drylands. The urbanization occurred mainly outside of the Ramsar designated site, but within the Ramsar site, there were increases in artificial wetland habitats and sea waters, with losses in natural wetland habitats. This study provides important monitoring information for managing the land resource in order to conserve the delta and its biodiversity in the future.

Freshwater wetlands are more threatened than any other ecosystem in the world. The current trend of freshwater wetland crisis arises a question 'whether it is conversion or conservation of wetlands' which is absolutely a dilemma. Therefore, this study aims critical evaluation at wetland governance and sustainability of urban wetlands in Colombo. The results reveal that the Colombo wetlands have lost their territory from a moderate to a severe degree. The wetland loss is higher in Kaduwela, Kolonnawa, and Kesbewa areas. Recently the causes of Colombo wetland loss has moved with a new frame. Land grabbing and green grabbing are the cycles that transforming the geostrategically important urban wetlands into hybrid ecosystems. Due to this, the frequency of flash flood events and inundation time has increased than ever. It is important to ensure urban freshwater wetland sustainability and surveillance following sustainable development goals and it is time to have freshwater ethics.

Groundwater in Tafilalet supports diverse faunas (stygofauna) that include many obligate groundwater-dependent species (stygobites). Prospecting campaigns on underground fauna in Tafilalt region were collected in a database that was later used to develop distribution patterns for each species harvested. Tafilalet is characterized

basins, wetlands are candidates to obviously display the radical change in water cycles related to anthropogenic activities. This chapter provides efficient methods to investigate water cycle modifications through the study of the water budgets within wetlands (Sidi El Hani Discharge Playa, Eastern Tunisia was taken as example). As concrete manifestation of this change, the dryness of in water springs or the decrease of their flows are concrete indicators of a real water crisis in Tunisia. In addition, salt consumption leads to a decreasing salinization.

Endoreic saline systems in Tunisia such as that of Chott El Guettar recorded the climatic changes as well as the setting of millennial cycles. This study deals with high-resolution proxies (colors variability, geochemistry, magnetic susceptibility, and grain size distribution) to find the different cycles. Cycles should be attributed either to internal or external forcing. The output of this study demonstrates a sun-driven Holocene cyclostratigraphy of 1000yr that was previously discussed in the literature. Since color studies were confirmed by direct analyses, this method was used to carry a higher resolution. The solar periodicity clearly identified with a continuous cycle of around 2500 years and with two others of around 1000 and 1660 years only during the Early Holocene. This work confirmed the cycles found by colors study through the real analyses such as geochemical and magnetic measurement.

High-resolution proxy-based paleoenvironmental records derived from wetlands provide important insights into climate changes over centennial to millennial timescales. This work aims to study the climatic cyclicity of the clayey fraction along a 78 cm core from the saline system of Mhabeul, located in Southeastern Tunisia. Based on the age model of Essefi et al., the core covers the last two millennia (»2100 yr). A high-resolution sampling of 2mm was carried out to obtain 380 samples evenly distributed along the core. In terms of analyses, the cumulative curve of the grain size distribution carried out by Fritsch laser apparatus percentages of the Clayey fraction underwent the spectral analysis. Based on the spectral analysis, millennial cycles of » 2500 yr, and 1000 yr are related to solar forcing. Solar irradiation is most likely responsible for the cyclic characteristics at 500 yr (600 yr) frequency.

Preface

Wetland are known globally as areas of swamps, natural water, permanent or temporary artificial waters where water is stagnant, in progress, melting or salty, including seawater areas no more than six meters deep at the lowest islands. Wetlands span about six percent of the Earth's surface (more than 2.5 million square kilometers) and currently have more than 2,400 Ramsar sites around the world. Wetlands are of the utmost importance: two thirds of the fish we consume depend on these areas at one moment or another and global food production depends on them (fish, shells, water and food for livestock, rice....). Wetlands can be considered a natural solution to the worldwide threat of Climate Change (CC). They are vital places for mankind and nature, basically because of the services and benefits provided by ecosystems at the social, economic, scientific, and educational levels as well as cultural recreational and aesthetic services for sustainable wishful development. which has become a determinant of modern times. In CO_2 absorption, coastal wetlands isolate and store carbon 55 times faster than tropical rainforests. So, it helps slow global warming and reduce pollution. In the face of extreme events (tsunamis, hurricanes, droughts, and floods), wetlands provide a protective barrier and strengthen resilience to CC. However, when wetlands are dried up and destroyed, they emit huge amounts of carbon. These ecosystems are known to have several serious imbalances due to drying for fear of insects or from the diversion of their water and the exploitation of their land in agriculture, construction, pollution caused by the dumping of solid and liquid waste, the use of insecticides and fertilizers, high salinity and the exploitation and conversion of land water. Wetlands are lost three times faster than forests and are among the most vulnerable ecosystems on Earth. In just 50 years—since 1970—35% of the world's wetlands have been lost.

The present book deals with the issue of environment generally and environmental significance of wetlands specifically. It's a contribution from us as researchers to raise national, regional, and global awareness about wetlands to achieve the rapid loss of land and to encourage action to preserve and restore it as part of achieving the Sustainable Development Goals (SDGs) particularly Goals 6, 14 and 15.

Climatic and Environmental Significance of Wetlands: Case Studies from Eurasia and North Africa produces innovative concepts, methodologies, tools, and applications for wetland biodiversity conservation, fresh water supply, agricultural production, food security, wetland management, and its impact on biodiversity. It assesses the cumulative risk posed to wetland habitats and species by human activities and explores the consequences for the delivery of ecosystem services (ES) and biodiversity at local, regional, and global scales, as well as the influence of CC on these ecosystems. Covering topics such as geochemistry, invasive species, and sedimentary change, this premier reference source is an indispensable resource for government officials, engineers, environmental managers, environmentalists, students and educators of higher education, researchers, and academicians.

This book opens with a chapter that situates the Monitoring of continental wetland dynamics and drivers of changes using Google Earth Engine. Wetlands around the world face anthropogenic and climate pressures that threaten key ecosystem processes such as the water cycle. Such threats are most acute for non-perennial systems that rely on occasional surface water flows to maintain important ecological functions. This work will focus on the use of remote sensing within a cloud-based computing platform to assess the spatio-temporal evolution of Dayet Aoua through the development of a water transition map between 1984 and 2019. It is also question of detecting and spatializing the main drivers of the various challenges, mainly the land use changes operated. Our study revealed that a combination of changes in climatic and land use change in the vicinity of Lake has recently affected the dynamic and the recovery capacity of Dayet Aoua. This study highlights the need to increase public awareness and decision-makers' knowledge concerning the relationship between land use change and wetlands degradation.

Chapter 2 highlights a case study of monitoring the land use and land cover in North Mediterranean wetlands. Monitoring Land Use / Land Cover changes is essential for planning and management activities to conserve a particular habitat. In this study, Land Use / Land Cover changes in the Gediz Delta (Turkey) between 1984-2020 was mapped using Earth Observation and Geographic Information Systems. The maps were built upon the Horizon-2020 Satellite-based Wetlands Observation Service processing methodology and algorithms. We compared changes inside and outside the Ramsar Area in the Gediz Delta. The results indicate more than 147% increase in built-up areas and decreases of 33% in natural wetland habitats and 27% in natural drylands. The urbanization occurred mainly outside of the Ramsar designated site, but within the Ramsar site, there were increases in artificial wetland habitats and sea waters, with losses in natural wetland habitats. This study provides important monitoring information for managing the land resource in order to conserve the Delta and its biodiversity in the future.

Chapter 3 reviews a local study aimed at a critical evaluation of wetland governance and sustainability of urban wetlands in Colombo, Sri Lanka. Its results revealed that the degree of loss of colombo wetlands ranged from very moderate to severe. Wetland losses are on the rise in in Kaduwela, Kolonnawa and Kesbewa areas. Recently the causes of Colombo wetland loss has moved with a new frame. Land grabbing and green grabbing are the cycles that transforming the geostrategically important urban wetlands into hybrid ecosystems. Due to this, the frequency of flash flood events and inundation time has increased than ever. It is important to ensure urban freshwater wetland sustainability and surveillance following sustainable development goals and it is time to have freshwater ethics.

Chapter 4 studied subterranean wetlands biodiversity in the context of climate change: Case study from Tafilalet, Southeastern Morocco. The study objectives are firstly, mapping the distribution of the stygobia fauna in Tafilalet region, secondly, to pursue the links between the specific richness of this fauna and global changes, and lately, to develop and test a risk assessment and decision-making framework for managing groundwater dependent ecosystems (GDEs) with declining water levels due to climate change, anthropogenic extraction, land use and land management. Based on this work and on the series of surveys conducted since the first sampling campaigns to date, several sites have been identified. Sites with a particular stylistic richness are called "hotspot". In such a conservation strategy, it is the "hot spots" that are given priority for preservation. To this end, a few guidelines that could ensure the sustainable conservation of this biodiversity are proposed.

Chapter 5 as part of the third section, 'Groundwater and Marine Biodiversity in a Context of Climate Change', highlights the Sustainability of Seafood Market under the Attack of Invasive Alien Species. This study aims to review the risk of invasive alien species in Turkish marines in the context of sustainable seafood market. Thus, this study investigates the challenges of invasive alien species on seafood market and marine biodiversity. By using, classifications, tables and figures, this study will determine the current profile of Turkish Seafood Market due to the risk level with Alien Species. The open access data obtained from Turkish Statistical Institute (TURKSTAT) and FAO (Food and Agriculture Organization of the United Nations) will be used to give main indicators for Turkish Seafood Market. Then, recent reports and news will be used to determine the efficiency of the implications and policies to fight against to Alien species in Turkey. By reviewing recent news and reports, it is thought to give an holistic approach for the related literature.

A section of this book brings together the case studies on hydrogeological and climatic aspects of Wetlands, past and current changes based on geochemical methods.

Chapter 6 investigates the water crisis in Tunisia during the Anthropocene and great acceleration wetlands as key sites of hydrogeological modifications. Due to overexploitation, anthropogenic pollution and climatic change during the

Anthropocene and Great Acceleration, Tunisia has already witnessed a severe water crisis. Since they are the natural outlets of their hydrological and hydrogeological basins, wetlands are more candidate to obviously display the radical change in water cycles related to anthropogenic activities. This chapter provides with efficient methods to investigate water cycle modifications through the study of the water budgets within wetlands (Sidi El Hani discharge playa, eastern Tunisia was taken as example). As concrete manifestation of this change, the dryness of in water springs or the decrease of their flows are concrete indicators of a real water crisis in Tunisia. In addition, salt consumption leads to a decreasing salinization.

Chapter 7 explored Geochemistry and sedimentology study of a core from Chott Geuttar, central Tunisia climatic and environmental significance. This study deals with high-resolution proxies (colors variability, geochemistry, magnetic susceptibility, and grain size distribution) to find the different cycles. Cycles should be attributed either to internal or external forcing. The output of this study demonstrates a sun-driven Holocene cyclostratigraphy of 1000yr that was previously discussed in the literature. Since color studies were confirmed by direct analyses, this method was used to carry a higher resolution. The solar periodicity clearly identified with a continuous cycle of around 2500 years and with two others of around 1000 and 1660 years only during the Early Holocene. This work confirmed the cycles found by colors study through the real analyses such as geochemical and magnetic measurement.

Chapter 8 focused on high-resolution proxy-based paleoenvironmental records derived from wetlands provide important insights into climate changes over centennial to millennial timescales. This work aims to study the climatic cyclicity of the clayey fraction along a 78 cm core from the saline system of Mhabeul, located in Southeastern Tunisia. Based on the age model of Essefi et al. (2014), the core covers the last two millennia (»2100 yr). A high-resolution sampling of 2mm was carried out to obtain 380 samples evenly distributed along the core. In terms of analyses, from the cumulative curve of the grain size distribution carried out by Fritsch laser apparatus percentages of the clayey fraction underwent the spectral analysis. Based on the spectral analysis, millennial cycles of »2500 yr, and 1000 yr are related to solar forcing. solar irradiation are most likely responsible for the cyclic characteristics at 500 yr (600 yr) frequency.

Abdelkrim Ben Salem
Faculty of sciences, Mohammed V University, Rabat, Morocco

Laila Rhazi
Faculty of sciences, Mohammed V University, Rabat, Morocco

Ahmed Karmaoui

Moulay Ismail University, Meknès, Morocco & Moroccan Center for Culture and Science, Morocco

REFERENCES

Essefi, E., Mefteh, S., Medhioub, M., & Yaich, C. (2014). Magnetic study of the heated and unheated sedimentary fillings of Sebkha Mhabeul, Southeast Tunisia: A geophysical method for paleoclimatic investigation and tephrochronological dating. *International Journal of Geophysics*. doi:10.1155/2014/908395

Acknowledgment

Wetlands provide a key service in ecosystem. Because of their ephemeral character and their small size, however, these vulnerable ecosystems are declining rapidly as climate change continues to surge and human activities expand. Rational management of wet ecosystems need accompanying actions covering research, systematic observation, and more.

The book introduces many innovative concepts, methodologies, tools, and applications for ecosystem service valuation, wetland biodiversity conservation, wetland management, and its impact on biodiversity through examples at the local level.

This book, after the grace of God Almighty, is the result of a collective effort in which many participated from the Mohammed V University in Rabat and Moulay Ismail University to all of them I owe thanks and appreciation, especially among them and the participants in editing. My co-editors Prof. Ahmed Karmaoui from the University of Moulay Ismail in Meknes, and Prof. Laila Rhazi from Mohammed V University in Rabat, for their wonderful effort in follow-up, proofreading, and revision. Big thanks go also to all the contributors who wrote the chapters from Turkey, Tunisia, Sri Lanka, France, and Morocco. Many thanks go to the reviewers to support us in the review process.

We appreciate the support of the IGI Global team, the encouragements and support of Prof. Mohammed Regragui, the Dean of the Faculty of Science (Mohammed V University in Rabat), the staff of the Botany and valuation of plant and fungal resources team and the Research Center Plant and Microbial Biotechnologies, Biodiversity, and Environment (BioBio), faculty of science, Rabat, Morocco. Many thanks go also to the whole Editorial Advisory Board, Prof. Ahmed El Aboudi, Prof. Said Moukrim, Prof. Issam Ifaadassan, Prof. Moustapha Arahou, And Prof. Hocein Bazairi.

Finally, my thanks go to my family members, my wife Hasna, my two children Walid and Aymane, my brothers and sisters and my parents for their patience and continued support.

This work was supported by BiodivRestore Program (RESPOND Project) funded by MESRSI (Ministry of Higher Education, Scientific Research, and Innovation of Morocco).

Editorial Advisory Board

Introduction

Wetlands are unique habitats that are flooded either permanently or seasonally during the year and are considered to be among the most diverse ecosystems, hosting an important and unique biodiversity of flora and fauna (Ramsar, 2007). These fragile ecosystems provide a large panorama of ecosystem services including water purification, flood control, carbon fixation, food and cultural benefits (Mitsch et al., 2015). Despite the importance of wetlands for both humans and wildlife, they continue to be among the most threatened ecosystems in the world (Brooks et al., 2002). Recent studies have shown that the surface area of natural Mediterranean wetlands has continued to decrease over the last decades with a significant decline in both water quality and quantity (Mediterranean Wetlands Observatory, 2018). The degradation is caused by different factors including urbanization, increases in intensive agriculture, invasive species and climate change (Leberger et al., 2020; Taylor et al., 2021). Climate change is and will continue to be one of the biggest threats to wetlands around the world (Erwin, 2009). It is imperative to address these threats in order to provide appropriate management solutions that will promote the sustainability of wetlands in the future. In order to share experiences and improve wetland conservation, we have compiled research articles from pilot sites from Eurasia to North Africa that focus on each of these threats through a different lens.

First, using a more socio-economical lens, we demonstrate wetland dynamics using monitoring information and socio-economic data from cloud-based computing in a case study from Morocco. This study revealed that a combination of changes in climatic and land use changes affected the dynamic and recovery capacity. Through another case study from Morocco, we provide an overview on how socio-economic pressures have affected natural wetlands, their biodiversity and ecosystem services over the last 30 years. This is further demonstrated in a remote sensing case study from Turkey where the land use changes over time are monitored and analyzed to determine the past and on-going threats impacting the wetlands.

The threat of urbanization for wetlands is addressed through a case study from Sri Lanka. Here we look into the ecology, governance and sustainability of urban wetlands. The results reveal that wetland losses have contributed to increases in frequency of flash flood events and inundation time.

Using a site from Morocco, we focus on groundwater and marine biodiversity. As in many wetlands located in arid climates, ground water is essential for supporting specific climate adapted species. This case aims to assess the species richness in the subterranean environment and then to develop and test a risk assessment and decision-making framework for managing groundwater-dependent ecosystems with declining water levels due to climate change, anthropogenic extraction, land use and land management. We then look into how invasive alien fish species are controlled in marine protected areas in Turkey. This study characterizes the current profile of the Turkish Seafood Market in relation to the risk level of alien species. The results demonstrate the efficiency of some actions and policies to fight alien species in Turkey.

Changes in hydrological and climatic conditions are further explored in the case study from Tunisia. This study investigates water cycle modifications through a perspective of water budgets within wetlands. Hydrological modifications are further studied in another case study from Tunisia, looking at the recorded climatic changes as well as the setting of millennial cycles. This work confirmed the usefulness of cycles found by color studies through the analyses of geochemical and magnetic measurements.

The articles and geographical territories covered in this book offer different perspectives and visions for addressing the climatic and environmental significance of wetlands. We acknowledge the importance of further research and information sharing on these topics in order to improve the conservation of wetlands in the future.

Lisa Ernoul
Tour du Valat, Research Institute for the Conservation of Mediterranean Wetlands, France

REFERENCES

Brooks, T. M., Mittermeier, R. A., Mittermeier, C. G., da Fonseca, G. A. B., Rylands, A. B., Konstant, W. R., Flick, P., Pilgrim, J., Oldfield, S., Magin, G., & Hilton-Taylor, C. (2002). Habitat loss and extinction in the hotspots of biodiversity. *Conservation Biology*, *16*(4), 909–923. doi:10.1046/j.1523-1739.2002.00530.x

Erwin, K. L. (2009). Wetlands and global climate change: The role of wetland restoration in a changing world. *Wetlands Ecology and Management*, *17*(1), 71–84. doi:10.100711273-008-9119-1

Leberger, R., Geijzendorffer, I. R., Gaget, E., Gwelmami, A., Galewski, T., Pereira, H. M., & Guerra, C. A. (2020). Mediterranean wetland conservation in the context of climate and land cover change. *Regional Environmental Change*, *20*(2), 67. doi:10.100710113-020-01655-0

Mediterranean Wetlands Observatory. (2018). *Mediterranean Wetlands Outlook 2: Solutions for sustainable Mediterranean Wetlands*. Author.

Mitsch, W., Bernal, B., & Hernandez, M. (2015). Ecosystem services of wetlands. *Int. J. Biodivers. Sci. Manag.*, *11*(1), 1–4. doi:10.1080/21513732.2015.1006250

Ramsar. (2007). *Le Manuel de la Convention de Ramsar: Guide de la convention sur les zones humides* (4th ed.). Ramsar.

Taylor, N. G., Grillas, P., Al Hreisha, H., Balkız, Ö., Borie, M., Boutron, O., Catita, A., Champagnon, J., Cherif, S., Çiçek, K., Costa, L. T., Dakki, M., Fois, M., Galewski, T., Galli, A., Georgiadis, N. M., Green, A. J., Hermoso, V., Kapedani, R., ... Sutherland, W. J. (2021). The future for Mediterranean wetlands: 50 key issues and 50 important conservation research questions. *Regional Environmental Change*, *21*(2), 33. doi:10.100710113-020-01743-1 PMID:33776560

Section 1
Wetland Monitoring

Chapter 1
Monitoring Continental Wetland Dynamics and Drivers of Changes Using Google Earth Engine

Said Moukrim
Research Center BIOBIO, Faculty of Sciences, Mohammed V University, Rabat, Morocco

Said Lahssini
https://orcid.org/0000-0003-1986-8240
National School of Forest Engineers, Morocco

Mouhssine Rhazi
Faculty of Sciences and Techniques, Moulay Ismail University, Morocco

Er-Riyahi Saber
Faculty of Letters and Human Sciences, Department of Geography, Moulay Ismail University, Morocco

Kamal Menzou
Water and Forests Department, Morocco

Mohammed El Madihi
Faculty of Sciences, Chouaib Doukkali University, El Jadida, Morocco & Research Center BIOBIO, Faculty of Sciences, Mohammed V University, Rabat, Morocco

Abdelkrim Ben Salem
https://orcid.org/0000-0002-2283-5928
Research Center BIOBIO, Faculty of Sciences, Mohammed V University, Rabat, Morocco

Abdelkader Benabou
Research Center BIOBIO, Faculty of Sciences, Mohammed V University, Rabat, Morocco

Hicham Mharzi-Alaoui
Mohammed VI Polytechnic University, Morocco

Issam Boukhris
University of Tuscia, Italy & CMCC Foundation, Italy

Laïla Rhazi
Research Center BIOBIO, Faculty of Sciences, Mohammed V University, Rabat, Morocco

DOI: 10.4018/978-1-7998-9289-2.ch001

ABSTRACT

Wetlands around the world face anthropogenic and climate pressures that threaten key ecosystem processes such as the water cycle. Such threats are most acute for non-perennial systems that rely on occasional surface water flows to maintain important ecological functions. This work will focus on the use of remote sensing within a cloud-based computing platform to assess the spatio-temporal evolution of Dayet Aoua through the development of a water transition map between 1984 and 2019. It is also question of detecting and spatializing the main drivers of the various challenges, mainly the land use changes operated. This study revealed that a combination of changes in climatic and land use change in the vicinity of the lake has recently affected the dynamic and the recovery capacity of Dayet Aoua. This study highlights the need to increase public awareness and decision-makers' knowledge concerning the relationship between land use change and wetlands degradation.

INTRODUCTION

Wetlands are rich and diverse and host nearly 10% of the world's biodiversity. They play several functions and provide various services to humans (Gardner and Finlayson, 2018). However, wetlands remain the most threatened ecosystems due to climate change and human activities (Dixon et al., 2016) and continue to disappear globally. Authors have investigated relations between surface water and climate variation or agriculture activities (Poudel et al., 2013). In Moroccan context, studies confirmed that the annual rainfall underwent significant downward trends and would certainly have a significant impact on wetlands (Sebbar, 2013). Also, emerging agriculture strategies promoted agriculture intensification even within arid lands. Intensive use of ground water has been subsideized (PMV, 2008). In mountainous lands such as Middle Atlas which was mainly devoted to extensive pastoral practices have been reoriented toward arboriculture with as consequence the development and the sinking of boreholes and pumping in the aquifers.

Area of the surface of water bodies have significant inter-annual and intra-annual variations. Unfortunately, due to resources lack, spatial and temporal information of surface water that is mandatory to guide conservation and management strategies remain scarce and a common challenge to scientist and managers in several regions mainly in low and middle-income countries.

To deal with the absence of an in-situ monitoring of the state of these ecosystems, earth observation offers opportunities for periodic and systematic monitoring of these areas and become an important data source for monitoring water surface dynamics. Open access satellite images, mainly acquired by Landsat, MODIS and Sentinel satellites, have opened up very interesting development perspectives with

Figure 1. Location of the Lake Dayet Aoua, Morocco

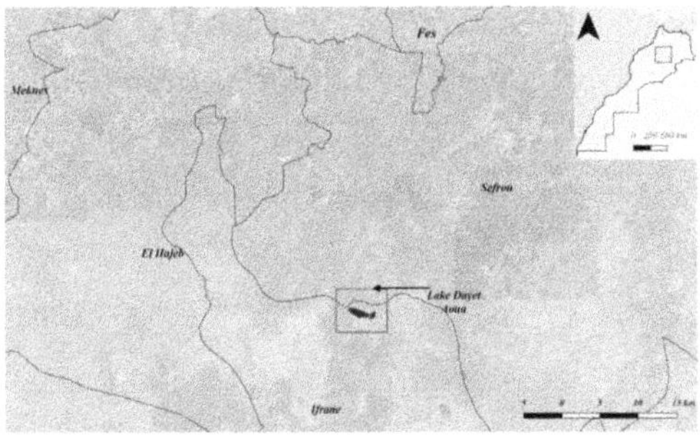

well-documented case studies for wetlands monitoring (Pekel et al., 2016). Indeed, specific indices were developped and are based on the physical characteristic of water relying on its lower reflection in near and shortwave infrared wavelength of the electormagnetic spectrum (Smith and Baker, 1981). The integration and processing of such data become possible with the recent development of cloud-based geospatial analysis platforms dedicated to spatial Big Data integrating both image archives from different providers, processing algorithms and distributed processing capabilities. As exemple, Google Earth Engine (GEE) is widely used.

Dayet Aoua, as natural lake of the Middle Atlas, is facing challenging issue during the last years with huge variability of its water surface and long periods of total drying up. Our study contribute to detect and to assess spatiotemporal evolution of *Dayet Aoua* water surface and the Land Use and Land-Cover (LULC) change between 1984 and 2019. Due to the absence of in situ data, authors focused on the use of remote sensing moderate-spatial-resolution images that will be analyzed within a cloud-based computing platform.

MATERIALS AND METHODS

Study Area

The lake *Dayet Aoua* is located in the eastern part of the Middle Atlas (northeast of Ifrane) at 1460 m (Figure 1) with an area of 150 ha, a width of 1100 m, a radius of 3500 m and a maximum volume of 2,5 million m^3 (Chillasse and Dakki, 2004). It is fed mainly by the river “Oued El Kantra” at times of high flow and occasionally by other water points with relatively lower flow and the waterbody surface varies according the seasons and the years with a maximum depth around 5 meter. The lake is surrounded by natural forests composed mainly of Cedar, holm oak and Pines in addition to collective lands and agricultural lands

In addition to the body of the lake, a surrounding area with a radius of 6 km to the lake centroid was defined to have a sufficiently broad vision of land occupation in the territory and thus to analyze the various transitions and changes in this land occupation between 1984 and 2019.

Datasets and Processing

The approach consists on the use of multiple datasets based on remote sensing images. Satellite images covering time frame steps 1984 to 2019 were classified and used to assess LULC. The Global Surface Water (GSW) dataset, developed by the European Commission's Joint Research Centre which capture the dynamism of surface water extent using Landsat imageries (Pekel et al., 2016), was used. For precipitations and drought, the SPEI database, which offers long-time, robust information about drought conditions at the global scale, with monthly time resolution was used.

The water transition map extracted from GSW provides information on the change in seasonality between the first (1984) and last years (2019) and captures changes between differents classes of water occurrence. These conversions refer to changes in state from the beginning and end of the observation record, they do not describe what happened in the intervening years. The Water Transition Map gives the changes between eleven classes of water occurrences (No Water, Permanent, New Permanent, Lost Permanent, Seasonal, New Seasonal, Lost Seasonal, Permanent to seasonal, Seasonal to Permanent, Ephemeral seasonal, Ephemeral Permanent).

The composite Landsat images for 1984 and Sentinel for 2019 were used. Images were selected based on the quality assessment band flags to minimize the effects of clouds and clouds' shadows. Simultaneously, the spectral bands were selected for classification. In addition, some vegetation indices, which are helpful for land use classification, were calculated and added to the bands. Also, the classification and accuracy assessment were performed. For result validation, some data are collected in the field and other samples were collected in Google Earth Pro for visual interpretation. Finally, the major land cover for 1984 and 2019 were mapped to identify the spatial pattern and process to understand the influence of LULC changes on the surrounding area of Lake *Dayet Aoua*.

All datasets were processed in Google Earth Engine (GEE). It's a cloud computing platform providing access to satellite images, geospatial data, tools and specific scripts (Gorelick et al., 2017; Patel et al., 2015). GEE relay on the use of complex algorithms, including "expert systems" which are an applied branch of artificial intelligence (AI), visual analytics, and evidential reasoning" (Gorelick et al., 2017).

Figure 2. Evolution of SPEI values between 1980 and 2018

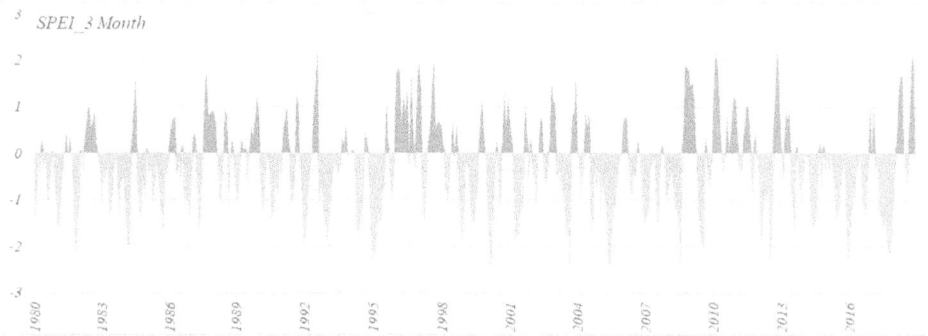

RESULTS AND DISCUSSION

Drought / Wet Conditions

The SPEI drought index provides results concerning the variability and trends of drought in the study area from 1981 to 2018 (Figure 2). The results indicate that SPEI time series at these time scales exhibit seasonal and inter-annual drought variability and shows that the episodes of droughts, different natures "moderate to extreme", are really present since 1980. It also shows an upward trend in the frequency of severe and extreme droughts which has increased considerably in recent decades. In another hand, it shows an increase in the frequency of severe, even extreme, wet conditions in some years.

These results corroborate the results of some studies that have highlighted the spatio-temporal evolution of annual rainfall in Morocco (Driouech, 2010; Sebbar, 2013). Such studies confirmed that from the 1970s, the annual rainfall totals underwent significant downward trends at most stations (including Ifrane which is close to the lake). Then, climate variation would certainly have a significant impact on the structure and functioning of many natural wetlands in the region (Grillas et al., 2021; Sebbar, 2013).

Inter and Intra-Annual Water Variations

During the study period, the lake has experienced periods of total drying out over a few months (from July to Nov-1995) (Figure 3a) or several months (from Sept-2000 to Dec-2002 or from Sept-2007 to Nov-2008) (Figures 3b and 3e). However, since jun-2018 to May-2019, *Dayet Aoua* has experienced a period of total drying out with small events of water accumulation on restricted areas. After this date, a

Figure 3. Images illustrate the surface areas of water bodies during and after different drying up period

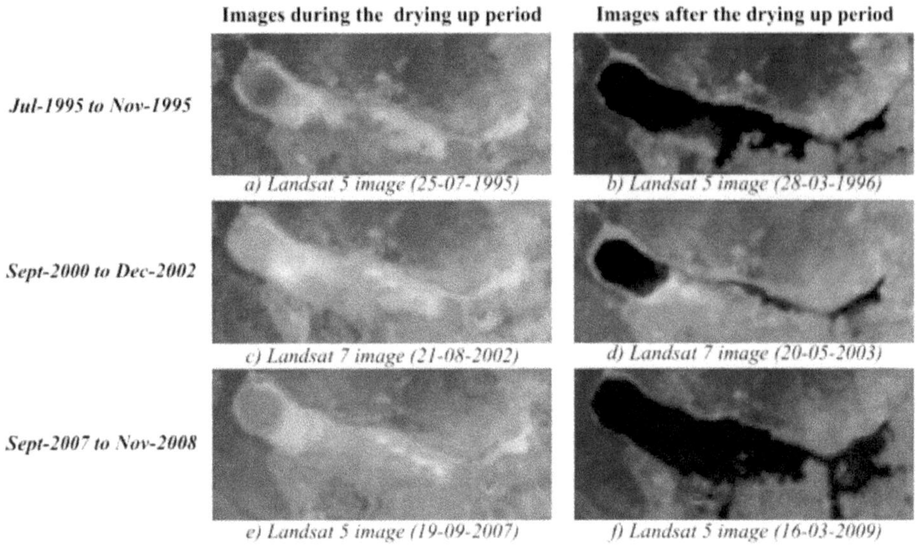

long period of total drying up began in June 2019 and continues until spring 2022. The satellite images highlighted those significant intra and inter-annual variations.

The analysis of the various satellite images made it possible to conclude that before 2010 the total dryings settled slowly, and the recovery of the lake settled quickly after each drying period (Figures 3b, 3d, and 3f). However, in recent years, authors noticed that period of total drying settled rapidly and continues over several years despite the fact that the year 2018 was an exceptional year in terms of rainfall (>1200 mm).

In concordance with our results, studies related to the representation of the fluctuation of the *Dayet Aoua* water level in relation with pluviometry (Sayad and Chakiri, 2010) have concluded also that the response of the lake to drought is a slower process after a succession of certain dry years; on the other hand, lake response to humidity is a rapid process and the lake regains an adequate level (lake recovery) of filling just after a good wet year. Unfortunately, this observation is no longer valid since 2017. The lake has not regained its adequate level with a new period of total drying up since the mid 2018 despite the fact that this year was exceptional in terms of rainfall (a record for precipitation). This situation means that other parameters have led to this condition.

Figure 4. Satellite images Show the submerged area of Dayet Aoua at: a) Apr-1984 & b) Dec-2019

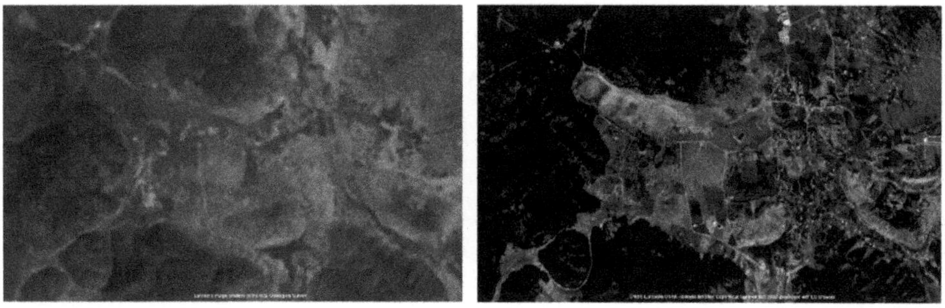

Water Class Transition

The use of the GEE platform facilitated analysis of the different images and maps (Figure 4) and enabled the development of the water transition table between the two dates (1984 and 2019). GEE help to create a grouped reducer for summing the area of each transition class within the study area, captures changes between different classes of water, and summarizes the area by transition class (Table. 1).

Table 1 shows the representativeness of each water transition classes which led us to calculate area of each class. The maximum extent of detected water over the period 1984-2019, which approximately concerns an area of 101 ha. The total water surface decreased significantly between the two dates selected.

In 2019, authors noted the absence of permanent waters. No transition from seasonal to permanent waters and no new permanent waters were recorded. Significant losses of permanent and seasonal water have been also recorded (concerning an

Table 1. Summary of transition class areas in Dayet Aoua between 1984 and 2019

Transition class areas	Percent (%)
Lost permanent	21.04
Seasonal	9.70
New seasonal	2.07
Lost seasonal	38.37
Permanent to seasonal	10.52
Ephemeral permanent	0.15
Ephemeral seasonal	18.15

Figure 5. Land use classification in 1984 and 2019 in a radius of 6 km at the centroid of the lake

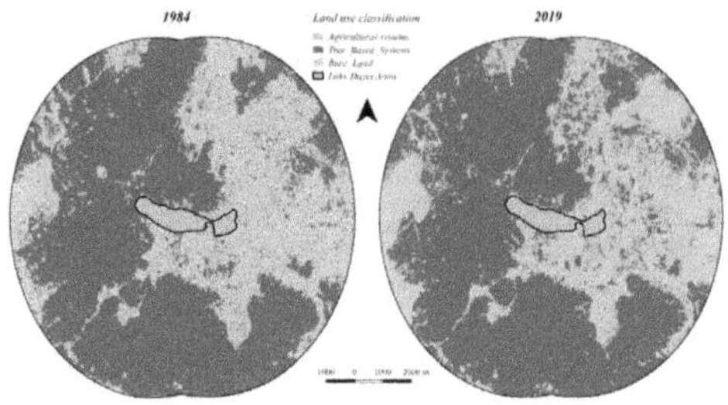

area of almost 60% of the Lake). Also, a transition from the permanent water class to the seasonal water class was recorded on 10.5%.

A combination of changes in climatic and socio-economics conditions has recently increased the stress on global resources (Alexander et al., 2015; Garnett et al., 2013), mainly excessive water exploitation in rural area. The growth of human population and its increasing consumption, the expansion of crops and the effects of climate change have intensified a variety of natural resource crises.

Land Use and Land Cover Changes

The surrounding area (a radius of 6 km) at the centroid of the lake allowed to have a sufficiently broad vision of the land use on the territory and to analyze the various transitions and changes in this land occupation between the two dates. The use of LULC information is essential to determine the environmental impacts of anthropogenic land-use and conversion in the study area.

The spatial analysis of the dynamics of land use (the LULC) (Figure 5) made it possible to conclude that between 1984 and 2019, the landscape dynamic reveals that all classes are affected by changes.

A significant number of pixels of each class have undergone conversions to other classes between 1984 and 2019. The most significant conversions remain those relating to the transitions from bare land to agricultural systems and from agricultural systems to tree based systems. Also, other conversions have been recorded from agricultural systems to bare land and tree-based systems to agricultural systems, which testifies to a long process of diffuse degradation of the emblematic ecosystems of the study area. Conversions to the tree-based systems concern transitions to permanent irrigated crops represented by arboriculture)

In recent years, with the change of land use in the surrounding areas, the water volume supplied to the Lake has been reduced, or even canceled in certain situations.

This reduction was accompanied by an exceptional demographic growth (47% between 1982 and 2014) (RGPH, 2014). This evolution has been accompanied by an increasing demand for drinking water, and industrial and agricultural use of water.

Irrigation development in general has had a strong impact on the water supply of *Dayet Aoua*. The development of forages and pumping in the aquifers, which are generally in hydraulic communication with the lakes, lowers their piezometric levels and consequently reduces the underground water flows supplying this lake. Also, the diversion of surface water for farms irrigation reduces the share of water supplying these lakes. Such trend was facilitated by an agricultural policy aiming to incentive irrigation and crop intensification with installation of arboriculture by promoting the shift from traditional to intensive agriculture (PMV, 2008). The drying up of this Lake confirms the trends observed at the national and regional scales and affects the innumerable ecosystem services offered by these spaces (Geijzendorffer et al., 2018).

CONCLUSION

This study analyzed mainly the spatial and temporal changes of Lake *Dayet Aoua* and LULC in the surrounding area. Our study revealed that a combination of changes in climatic and socio-economics conditions has recently affected the dynamic, extent and the recovery capacity of the Lake.

The results reveal that over the past four decades, the recurrence of unfavorable climatic conditions and LULC in the vicinity of Lake *Dayet Aoua* already show an impact on this ecosystem by impacting its recovery capacity. Thus, there is an urgent need to increase public awareness and decision-makers' knowledge concerning the effect of the relationship between LULC, water resources and natural wetland degradation.

Also, the monitoring framework used will contribute to the evaluation of the achievement of the objectives of the national strategy for wetlands 2015-2024 and its variations relating to the rational and sustainable use, and a resilient supply of ES in the benefit of local beneficiary populations.

ACKNOWLEDGMENT

This work was supported by BiodivRestore Program (RESPOND Project) funded by MESRSI (Ministry of Higher Education, Scientific Research and Innovation of Morocco).

REFERENCES

Alexander, P., Rounsevell, M. D., Dislich, C., Dodson, J. R., Engström, K., & Moran, D. (2015). Drivers for global agricultural land use change: The nexus of diet, population, yield and bioenergy. *Global Environmental Change*, *35*, 138–147. doi:10.1016/j.gloenvcha.2015.08.011

Chillasse, L., & Dakki, M. (2004). Potentialités et statuts de conservation des zones humides du Moyen-Atlas (Maroc), avec référence aux influences de la sécheresse. [Potentialities and conservation status of wetlands in the Middle Atlas (Morocco), with reference to the influences of drought.] *Science et changements planétaires/ Sécheresse, 15*(4), 337-345.

Dixon, M. J. R., Loh, J., Davidson, N. C., Beltrame, C., Freeman, R., & Walpole, M. (2016). Tracking global change in ecosystem area: The Wetland Extent Trends index. *Biological Conservation*, *193*, 27–35. doi:10.1016/j.biocon.2015.10.023

Driouech, F. (2010). *Distribution des précipitations hivernales sur le Maroc dans le cadre d'un changement climatique: descente d'échelle et incertitudes.* [*Distribution of winter precipitation over Morocco in the context of climate change: downscaling and uncertainties.*] [Doctoral dissertation, Universite de Toulouse, France].

Gardner, R. C., & Finlayson, M. (2018). Global wetland outlook: State of the world's wetlands and their services to people 2018. *Secretariat of the Ramsar Convention.*

Garnett, T., Appleby, M. C., Balmford, A., Bateman, I. J., Benton, T. G., Bloomer, P., Burlingame, B., Dawkins, M., Dolan, L., Fraser, D., Herrero, M., Hoffmann, I., Smith, P., Thornton, P. K., Toulmin, C., Vermeulen, S. J., & Godfray, H. C. J. (2013). Sustainable intensification in agriculture: Premises and policies. *Science*, *341*(6141), 33–34. doi:10.1126cience.1234485 PMID:23828927

Geijzendorffer, I., Chazée, L., Gaget, E., Galewski, T., Guelmami, A., Perennou, C., & McInnes, R. (2018). Mediterranean wetlands outlook 2: Solutions for sustainable Mediterranean Wetlands. *Secretariat of the Ramsar Convention.*

Gorelick, N., Hancher, M., Dixon, M., Ilyushchenko, S., Thau, D., & Moore, R. (2017). Google Earth Engine: Planetary-scale geospatial analysis for everyone. *Remote Sensing of Environment*, *202*, 18–27. doi:10.1016/j.rse.2017.06.031

Grillas, P., Rhazi, L., Lefebvre, G., El Madihi, M., & Poulin, B. (2021). Foreseen impact of climate change on temporary ponds located along a latitudinal gradient in Morocco. *Inland Waters*, *11*(4), 492–507. doi:10.1080/20442041.2021.1962688

Patel, N. N., Angiuli, E., Gamba, P., Gaughan, A., Lisini, G., Stevens, F. R., Tatem, A. J., & Trianni, G. (2015). Multitemporal settlement and population mapping from Landsat using Google Earth Engine. *International Journal of Applied Earth Observation and Geoinformation, 35*, 199–208. doi:10.1016/j.jag.2014.09.005

Pekel, J. F., Cottam, A., Gorelick, N., & Belward, A. S. (2016). High-resolution mapping of global surface water and its long-term changes. *Nature, 540*(7633), 418–422. doi:10.1038/nature20584 PMID:27926733

PMV. (2008). *Plan Maroc Vert* [*Green Morocco map*]. MAPMDREF.

Poudel, D. D., Lee, T., Srinivasan, R., Abbaspour, K., & Jeong, C. Y. (2013). Assessment of seasonal and spatial variation of surface water quality, identification of factors associated with water quality variability, and the modeling of critical nonpoint source pollution areas in an agricultural watershed. *Journal of Soil and Water Conservation, 68*(3), 155–171. doi:10.2489/jswc.68.3.155

Sayad, A., & Chakiri, S. (2010). Impact de l'évolution du climat sur le niveau de Dayet Aoua dans le Moyen Atlas marocain. [Impact of climate change on the level of Dayet Aoua in the Moroccan Middle Atlas.] *Science et changements planétaires/ Sécheresse, 21*(4), 245-251.

Sebbar, A. (2013). Etude de la variabilité et de l'évolution de la pluviométrie au Maroc (1935-2005): Réactualisation de la carte des précipitations. [Study of the variability and evolution of rainfall in Morocco (1935-2005): Updating of the rainfall map.] [Doctoral Dissertation, University of Hassan II Mohammedia, Turkey].

Smith, R. C., & Baker, K. S. (1981). Optical properties of the clearest natural waters (200–800 nm). *Applied Optics, 20*(2), 177–184. doi:10.1364/AO.20.000177 PMID:20309088

Chapter 2
Assessing the Impacts of Land Use and Land Cover Changes 1984-2020 on Wetland Habitats in the Gediz Delta (Turkey)

Anis Guelmami
Tour du Valat, Research Institute for the Conservation of Mediterranean Wetlands, France

Dilara Arslan
Tour du Valat, Research Institute for the Conservation of Mediterranean Wetlands, France

Lisa Ernoul
Tour du Valat, Research Institute for the Conservation of Mediterranean Wetlands, France

ABSTRACT

Monitoring land use / land cover changes is essential for planning and management activities to conserve a particular habitat. In this study, the authors mapped land use / land cover changes in the Gediz Delta (Turkey) between 1984-2020 using Earth Observation and Geographic Information Systems. The maps were built upon the Horizon-2020 satellite-based wetlands observation service processing methodology and algorithms. The authors compared changes inside and outside the Ramsar Area in the Gediz Delta. The results indicate more than 147% increase in built-up areas and decreases of 33% in natural wetland habitats and 27% in natural drylands. The urbanization occurred mainly outside of the Ramsar designated site, but within the Ramsar site, there were increases in artificial wetland habitats and sea waters, with losses in natural wetland habitats. This study provides important monitoring information for managing the land resource in order to conserve the delta and its biodiversity in the future.

DOI: 10.4018/978-1-7998-9289-2.ch002

INTRODUCTION

Wetlands are an essential part of human well-being and play an integral role for many organisms to complete their life cycles (Mitsch and Gosselink, 2000; Ramsar Convention Secretariat, 2016). They provide valuable ecosystem services to people such as provisioning water and food, flood regulation, erosion prevention, groundwater discharge, water purification, nutrient recycling, carbon storage, and recreational activities (Ramsar Convention Secretariat 2016). Despite their vital importance, wetlands have been one of the most transformed ecosystems over the last hundred years, with a loss of 87% of the world's wetlands since the 1700s. This trend increased four-fold in the 19th century (Davidson, 2014; Ramsar Convention on Wetlands, 2018). Agricultural extension, creation of artificial wetlands and urban development have been the main drivers of natural wetland loss and degradation over the last 200 years and continue to threaten the remaining 12.1 million km^2 worldwide (Zedler and Kercher, 2005; Hu et al., 2017; Ramsar Convention on Wetlands, 2018). It is estimated that approximately 35% of wetlands have been disturbed or irreversibly destroyed after the 1970s (Ramsar Convention on Wetlands, 2018). On the other hand, human-made wetlands (such as dams and reservoirs, salinas, rice fields, etc.) have increased since the 1970s (and earlier) and as of 2018 they constitute 12% of existing wetlands around the world (Ramsar Convention on Wetlands, 2018). In the Mediterranean Basin, natural wetlands have decreased by 48% since the 1970s, mainly through the conversion into agricultural lands, human-made wetlands and built-up areas (MWO, 2018). The human-made wetlands have been shown to host an important biodiversity and play a key role for waterbirds, but the changes in water supplies and management for these new wetlands are often a detriment to natural wetlands (Ramsar Convention on Wetlands, 2018; MWO, 2018).

There are several important international conventions that aim to protect wetlands including the Ramsar Convention, the Convention on International Trade in Endangered Species, the Convention on Biological Diversity, the Convention on Migratory Species and the Agreement on the Conservation of African-Eurasian Migratory Waterbirds (Christoffersen, 1997). Following this trend, Turkey has also showed the will to protect its wetlands at the international level by signing the Ramsar Convention in 1994 and at the national level by issuing a "Regulation for Protection of the Wetlands" in 2002. This has resulted in the designation of 14 wetlands as Ramsar sites and 56 that are protected by national laws (GDNCNP, 2020). Although this is a good step forward, it represents a much lower percentage of designation compared to other Mediterranean countries (for example there are more than 70 Ramsar sites in Spain, 50 in Algeria and 35 in mainland France). It is also important to note that despite these measures, wetlands in Turkey continue to be threatened by human activities such as intensive agriculture, urbanization, illegal poaching, pesticide use and hydrological modifications (Nilguuml, Alpay, & Emel, 2009; Scaramelli, 2020). Between 1920 and 1980, wetlands in Turkey were massively dried to acquire new agricultural lands, create settlements and combat water-related diseases such as malaria (Nilguuml, Alpay, & Emel, 2009; Scaramelli, 2020). As a consequence, almost 300,000 hectares of natural wetlands have been degraded or destroyed over the last century (Ataol and Onmuş, 2021). Furthermore, due to water abstraction policies implemented during the last decades in Turkey, with massive investments in dam construction (mainly for irrigation), many “untouched” natural wetlands are receiving less and less surface water, transforming their permanent water regimes into temporarily flooded zones, or even disappearing completely, as in the case of Lake Akşehir (Perennou et al., 2018).

One of the most important wetlands in Turkey is the Gediz Delta, located on the Aegean coast. In 1998, the site was classified as a Wetland of International Importance according to the Ramsar Convention due to its significant role in waterbird breeding and migration (Gediz Delta Management Plan, 2007). The Delta also hosts 14 freshwater fish species, 35 herptile species, and 8 mammal species (Gediz Delta Management Plan, 2007; Arslan et al., 2018). In addition to being a Ramsar site, the Delta has also been declared a National Wildlife Protection Area since 1982 and a Natural Protected Area since 1985. Despite these cumulated protection labels, the Gediz Delta has been severely impacted by human activities for more than a century and continues to be threatened by new projects (Avdan, 2020; Arslan et al., 2021).

Gediz Delta is located close to İzmir city, the third-largest city in the country. With a total population estimated at almost 5 million inhabitants and an annual growth rate of 10.8%, the population in Izmir has increased by approximately 120% since the 1980s (TUIK, 2020). This population growth has led to a drastic increase in the demand for natural resources, especially land for new urban settlements and crops and freshwater for irrigation, domestic and industrial uses. Due to the geographic proximity of the Delta to the city, the demographic pressure has increased the intensity of threats to the natural habitats and the biodiversity (Sıkı, 2020; Avdan, 2020; Arslan et al., 2021). Furthermore, the course of the Gediz River and its tributaries have been highly modified through the construction of dams and channels, with the most important changes occurring during the 1880s, when the riverbed was redirected from the south to the north of the Delta (Erinç, 1955; Aksu et al., 1990; Eken, 1997; Gediz Delta Management Plan, 2007). As a result, the Delta has lost some of its important habitats including the Çigli Marshes (140 ha of shallow freshwater marshes located in the south of the Delta) (Eken, 1997).

There have been some previous studies to understand these changes within the strictly protected area based on earth observation (EO) data (see for example Ernoul et al., 2012; Bolca et al., 2014). However, these studies were limited in their geographic scope and did not address the threats based on a systemic approach comparing land use / land cover (LULC) changes inside and outside of the Ramsar site. Thus, the current study aims to use freely available EO-based data and time series to fill this knowledge gap.

In this study, we used remote sensing approaches to assess the habitat dynamics and LULC changes in the Gediz Delta, based on a hybrid classification system combining CORINE Land Cover with Ramsar definitions (Mediterranean Wetland Observatory, 2014; Perennou et al., 2018) over a 36 year time period (1984-2020). The main objectives were to (1) assess the main spatial-temporal LULC changes during the period from 1984 to 2020, (2) quantify their impacts on natural wetland habitats and (3) evaluate the difference in and outside of the Ramsar designated area. This monitoring technique has been standardized (Weise et al., 2020) and can be replicated to other wetlands and/or basins in Turkey or around the Mediterranean Basin to monitor the evaluation of wetlands on a larger scale.

METHODOLOGY

Study Area

Gediz Delta (38þ 30'N, 26þ 55'E) is one of the most important wetlands in the Mediterranean Basin and is located on the Aegean coast in the west of Turkey. It extends over approximately 40,000 ha, including 14,900 ha designated as a Wetland of International Importance under the Ramsar Convention and covering the provinces of Uşak, Manisa, and Izmir, starting from the Gediz River in the province of Kütahya and formed by aggregations of the alluvial materials carried by the Gediz River and flowing into the Aegean Sea just north of Izmir (Arslan et al., 2018). Gediz Delta is composed of a mosaic of salt and freshwater marshes, saltpans, four main lagoons (Homa, Çilazmak, Kırdeniz and Taş), dry hilly habitats and also includes agricultural lands and urban settlements (Gediz Delta Management Plan, 2007). The study area includes not only the Ramsar site, but also takes into account the area between the old and new riverbeds of the Gediz River and the lower Gediz.

Mapping land use / land cover (LULC) in the Delta and assessing changes over time

Land use / land cover (LULC) maps were produced within the study area using Landsat TM, Landsat ETM and Sentinel-2 time series covering 1984, 1990, 2000, 2010 and 2020 annual periods. Based on these maps, it was then possible to analyze the long-term LULC dynamics between 1984 and 2020 and to derive status and trend statistics and indicators corresponding to each habitat class or group of aggregated habitat classes.

The mapping approach was built upon the satellite-based wetlands observation service (SWOS) image and data processing methodology (Weise et al., 2020). It includes object and rule-based classification algorithms and integrates a hybrid LULC nomenclature combining CORINE Land Cover (CLC) classes with the Ramsar wetland habitat definitions (Mediterranean Wetland Observatory, 2014; Perennou et al., 2018). Each annual LULC map captures the status of wetland ecosystems regarding their habitat composition, their spatial extent and the main agriculture and urbanization related pressures they face. By comparing and crossing these annual LULC maps together, it was also possible to analyze trends over time (losses, gains and conversion types for each LULC class). In order to facilitate the discussion of the obtained results, all the defined LULC classes were grouped into 6 main categories (Perennou et al., 2018): "Built up areas" (includes also commercial and transport infrastructures, mines, construction sites and urban green areas), "Agricultural areas" (excluding rice fields and wet meadows), "Natural Drylands" (e.g. forests, shrubland, natural grasslands, etc.), "Natural wetlands" (based on Ramsar definitions and including wet meadows), "Artificial Wetlands" (based on Ramsar definitions and including rice fields) and "Sea". These categories were used to build the status and trend indicators which have been calculated for the entire mapped area.

RESULTS

LULC Trends

The LULC classification results for the Gediz Delta, with aggregated classes, are summarized in Figure 1 and their trends can be seen in Figure 2. In 1984, more than 19% the of the mapped area was covered by natural wetland habitats. This area decreased to 11% in 36 years, representing a net loss of 8,820 ha (with an average declining rate of 3% per year). Even if these habitats represent a higher percentage of land cover within the limits of the Ramsar site (more than 52% in 1984), the trend is similar with a loss of 32% between 1984 and 2020 (while it was 33% for the entire Delta). One specific natural wetland habitat that underwent important changes over the studied period was reedbeds, with a net loss of 47% between 1984 and 2020 within the Gediz Ramsar site. Most of the decline occurred before 2000.

Most of the natural dryland habitats, which mainly include forests, shrubland and grasslands, are located outside of the Ramsar site (Figure 1) and have also decreased from 21% in 1984 within the mapped area to 15% in 2020. Agriculture surfaces remain relatively stable, even showing a slight increase from 34% to 37%

(a net gain of 2,333 ha, from 28,189 ha to 30,522 ha, with a maximum of 31,775 ha in 1990). The highest rate of increase between 1984 and 2020, were built-up areas (including commercial and transport infrastructures, mines, construction sites and green urban areas), representing a growth of 147% in the Delta. However, urban sprawl is much less important within the limits of the Ramsar site, with only 16% increase during the monitored 36-year time period (from 647 ha to 748 ha). Lastly, sea and marine habitats have also increased by 11% within the mapped area, from 13,629 ha in 1984 to 15,148 ha in 2020. This increasing rate of marine areas is almost three times higher within the Ramsar site limits (31% between 1984 and 2020).

Main LULC Driving Factors of Change

According to the LULC maps and indicators obtained here, the main driving factor for change at the scale of the Gediz Delta is urban sprawl. Most of the observed land use changes were conversions into built-up areas. This LULC change impacted more than 6,482 ha between 1984 and 2020, which is the largest change in land use in comparison with the five other categories. However, if we look at the trends within the Ramsar site, the most important changes are the creation of new artificial wetland habitats (1,387 ha of the observed transformation), followed by the rapid progression of the sea and marine waters (1,158 ha of the observed transformation), especially in the southern part of the Delta. These changes were observed at the detriment of natural wetland habitats for both the Ramsar site and at the scale of the Gediz Delta (Figure 3). On the other hand, the expansion of built-up areas has been primarily at the expense of natural drylands and agriculture areas (both within and outside of the Ramsar site) and has not greatly impacted natural wetlands.

DISCUSSION

Remote sensing was used to monitor the Land Use / Land Cover changes in the Gediz Delta over a 36-year time period. This monitoring quantified an increase in built-up areas (+147%) in the Gediz Delta between 1984 and 2020. In addition, the maps clearly showed a significant regression of natural wetland habitats (33%) and natural drylands (27%) in the Gediz Delta during the same period. The main drivers of change for the loss of natural wetland habitats and peripheral forest, shrubland and grassland habitats were urban expansion and drainage to combat malaria (Şekercioğlu et al., 2011). Despite the protection status and development of a management plan, the surface area of natural wetlands in the Delta are very limited, and urbanization and agricultural pressures continue to threaten them (Bolca et al., 2014).

Figure 1. Location of the mapped area (the Gediz Delta) and main LULC results for 1984 (a); 1990 (b); 2000 (c); 2010 (d) and 2020 (e) annual time periods

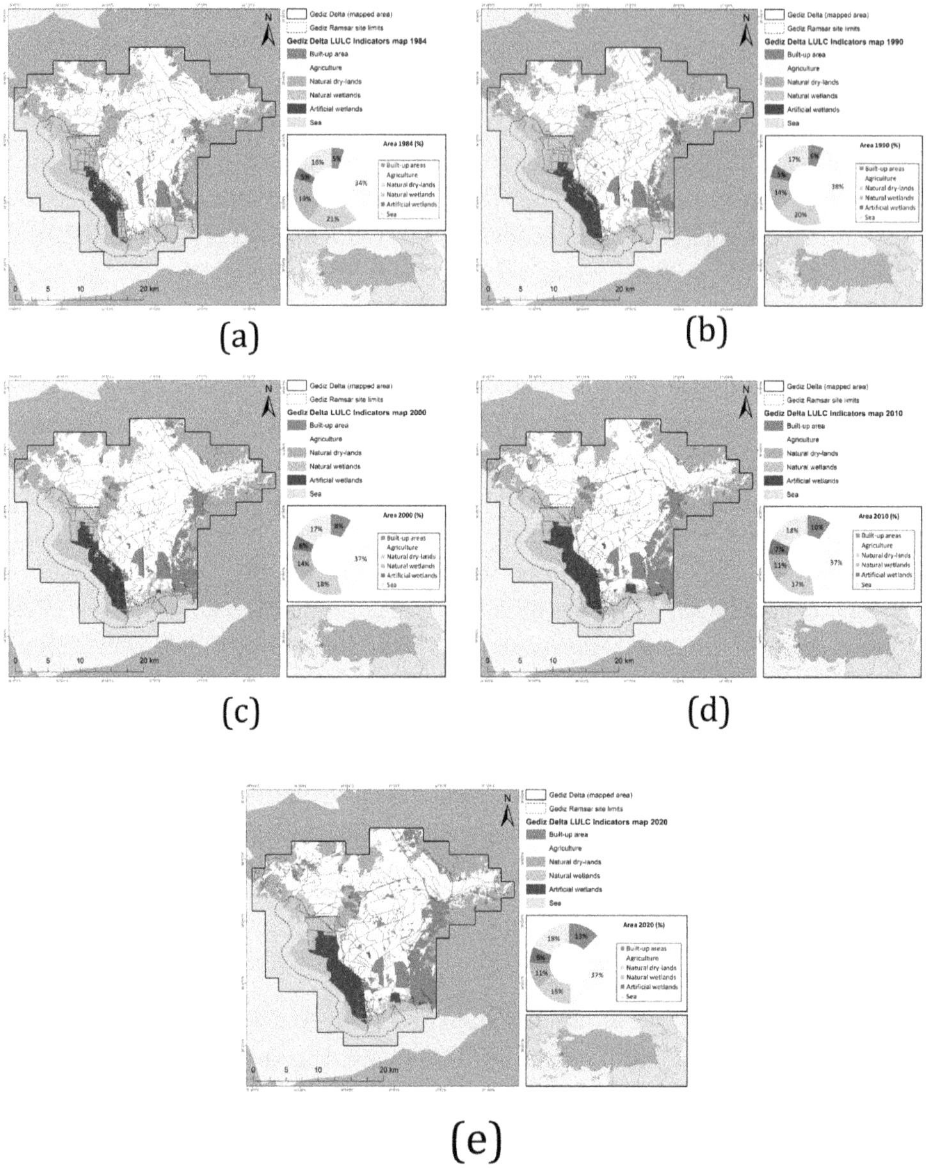

Another important threat to the Delta's natural wetland habitats is their significant conversion into ponds for salt production. The surface area of saltpans increased by 63% between 1984-2020. Another important change was the loss of almost half area of the existing reedbeds between 1984 and 2020. However, since 2019 a conservation

Figure 2. Main trends of the aggregated LULC categories (in ha) for five annual time periods between1984-2020 at the scale of the Gediz Delta (a) and within the limits of the Ramsar site (b)

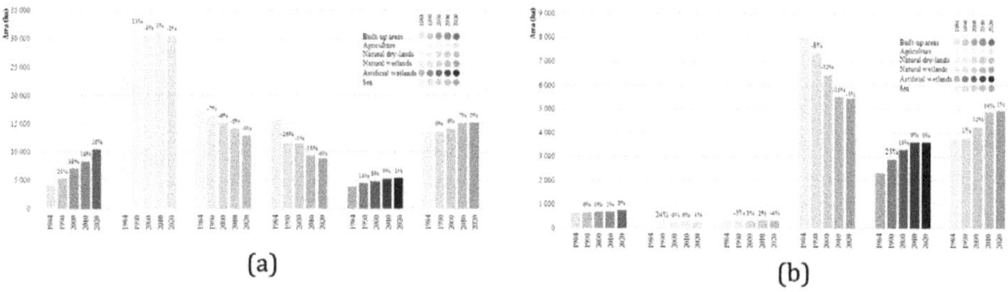

measure of this habitat was adopted, through the pumping of freshwaters using a system of canals (Avdan, 2020). Nonetheless, the constraints of limited freshwater resources caused by climate change, agricultural intensification and new dams risk to put additional pressures on the natural wetlands in the Delta in the future. Other minor trends were also observed with the transformation of ancient built-up or agricultural areas into natural dry- or wet-lands, or even artificial wetlands. According to the maps and derived change indicators, they are mainly due to the conversion of some cultivated croplands into wet meadows for instance (considered

Figure 3. LULC categories in 2020 classified according to their origins in 1984 and expressed in percentage for the Gediz Delta (a) and within the limits of the Ramsar site (b)

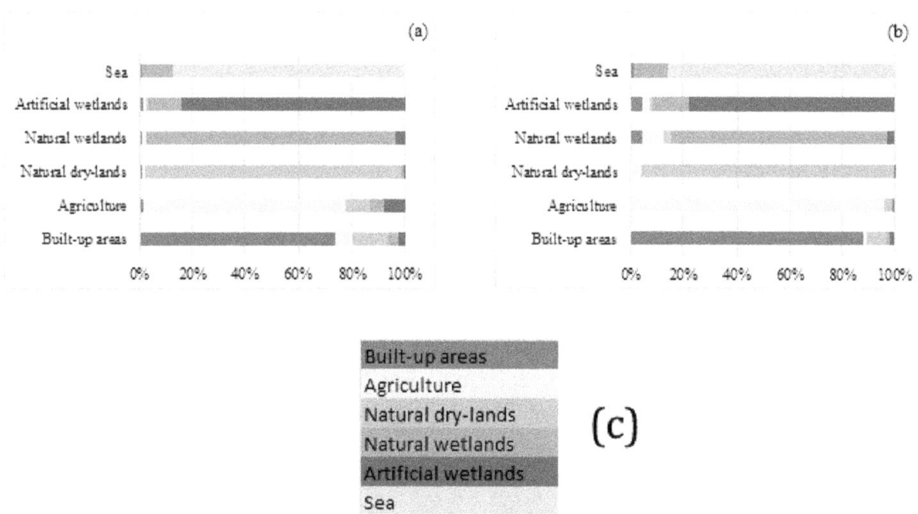

here as natural wetlands), or the regeneration of natural habitats after the end of an extraction activity (e.g. old sand or gravel pits).

The trends that we observed in the Gediz Delta are consistent with trends around the world. There are more important losses in natural wetland habitats as compared to other terrestrial habitats around the world (Ramsar Convention Secretariat, 2018). These trends are also similar to those observed in the Mediterranean Basin (Mediterranean Wetland Observatory, 2018) and greatly impact the biodiversity living in and around the wetlands (Galewski et al., 2011). The continued monitoring of wetland evolution both through remote sensing and fieldwork is urgent to advocate for better conservation management and to reduce the on-going threats to these essential ecosystems.

CONCLUSION

Wetlands habitats are considered to be some of the most threatened ecosystems worldwide (Finlayson et al., 1999; Geijzendorffer et al., 2018; Davidson et al., 2018) and the geographic proximity of the Gediz Delta to Izmir further increases the threats. Despite current protection laws and public awareness of the importance of this area, this study highlighted that conversion of natural wetlands, natural drylands and extensive agricultural landscapes to urban landscapes has increased in the last decades. These habitats are often outside of the Ramsar site, yet they offer important breeding and feeding habitats for many species (Galewski and Devictor, 2016) and many valuable ecosystem services to the society (MWO, 2018). Environmental heterogeneity is regarded as one of the most important components to sustain high biodiversity (Stein et al., 2014), therefore, we highly recommend that conservation activities include all of the Delta's natural ecosystems, especially those largely forgotten in conservation policies.

REFERENCES

Aksu, A. E., Konuk, T., Ulug, A., & Duman, M., & PIPER, D. W. (1990). Quaternary tectonic and sedimentary history of Eastern Aegean Sea shelf area. *Jeofizik*, *4*(1), 3–35.

Arslan, D., Çiçek, K., Döndüren, Ö., & Ernoul, L. (2021). Threat Ranking to Improve Conservation Planning: An Example from the Gediz Delta, Turkey. *Land (Basel)*, *10*(12), 1381. doi:10.3390/land10121381

Arslan, D., Oliver, A., Yaşar, Ç., Ismail, İ. B., Döndüren, Ö., Ernoul, L., ... Çiçek, K. (2018). Distribution and current status of herpetofauna in the Gediz Delta (Western Anatolia, Turkey). *Herpetology Notes*, *11*, 1–15.

Ataol, M., & Onmuş, O. (2021). Wetland loss in Turkey over a hundred years: Implications for conservation and management. *Ecosystem Health and Sustainability*, *7*(1), 1930587. doi:10.1080/20964129.2021.1930587

Avdan, A. O. (2020). *The making of consurbia: Conservation, urbanization, and socio-environmental change in Turkey's Gediz Delta.* [Thesis, Simon Fraser University, Canada].

Bolca, M., Özen, F., & Güneş, A. (2014). Land use changes in Gediz Delta (Turkey) and their negative impacts on wetland habitats. *Journal of Coastal Research*, *30*, 756–764.

Christoffersen, L.E. (1997). IUCN: A Bridge-Builder for Nature Conservation. *Green Globe Yearbook*, 59–70

Davidson, N. C. (2014). How much wetland has the world lost? Long-term and recent trends in global wetland area. *Marine and Freshwater Research*, *65*(10), 934–941. doi:10.1071/MF14173

Davidson, N. C., Fluet-Chouinard, E., & Finlayson, C. M. (2018). Global extent and distribution of wetlands: Trends and issues. *Marine and Freshwater Research*, *69*(4), 620–627. doi:10.1071/MF17019

Eken, G. (1997). The breeding population of some species of waterbirds at Gediz Delta, Western Turkey. *Zoology in the Middle East*, *14*(1), 53–68. doi:10.1080/09397140.1997.10637704

Erinç, S. (1955). *Gediz ve Büyük Menderes deltalarının morfolojisi* [Morphology of Gediz and Great Menderes deltas]. Türk Coğrafya Kur. Yay.

Ernoul, L., Sandoz, A., & Fellague, A. (2012). The evolution of two great Mediterranean Deltas: Remote sensing to visualize the evolution of habitats and land use in the Gediz and Rhone Deltas. *Ocean and Coastal Management*, *69*, 111–117. doi:10.1016/j.ocecoaman.2012.07.026

Finlayson, C. M., Davidson, N. C., Spiers, A. G., & Stevenson, N. J. (1999). Global wetland inventory–current status and future priorities. *Marine and Freshwater Research*, *50*(8), 717–727. doi:10.1071/MF99098

Galewski, T., Collen, B., McRae, L., Loh, J., Grillas, P., Gauthier-Clerc, M., & Devictor, V. (2011). Long-term trends in the abundance of Mediterranean wetland vertebrates: from global recovery to localized declines. *Biological Conservation.*

Galewski, T., & Devictor, V. (2016). When common birds became rare: Historical records shed light on long-term responses of bird communities to global change in the largest wetland of France. *PLoS One, 11*(11), e0165542.GDNCNP (2020)

Korunan Alanlar İstatistikleri (2019). *Protected Area Statistics.* [Data set]. Turkey Ministry of Agriculture and Forestry. https://www.tarimorman.gov.tr/DKMP/Menu/18/Korunan-Alan-Ista
tistikleri#>.

Gediz Delta Management Plan. (2007). *Gediz Deltası Sulak Alan Yönetim Planı* [Gediz Delta Management Plan].

Geijzendorffer, I., Chazée, L., Gaget, E., Galewski, T., Guelmami, A., Perennou, C., & McInnes, R. (2018). *Mediterranean wetlands outlook 2: Solutions for sustainable Mediterranean Wetlands.* Secretariat of the Ramsar Convention.

Hu, S., Niu, Z., Chen, Y., Li, L., & Zhang, H. (2017). Global wetlands: Potential distribution, wetland loss, and status. *The Science of the Total Environment, 586*, 319–327. doi:10.1016/j.scitotenv.2017.02.001 PMID:28190574

Nilguuml; l, K., Alpay, T., & Emel, B. (2009). Wetland management in Turkey: Problems, achievements and perspectives. *African Journal of Agricultural Research, 4*(11), 1106-1119.

Mediterranean Wetland Observatory. (2014). *Land cover: Spatial dynamics in Metierranean coastal wetlands from 1975 to 2005.Thematic collection 2*. Tour du Valat.

Mitsch, W. J., & Gosselink, J. G. (2000). The value of wetlands: Importance of scale and landscape setting. *Ecological Economics, 35*(1), 25–33. doi:10.1016/S0921-8009(00)00165-8

Perennou, C., Guelmami, A., Paganini, M., Philipson, P., Poulin, B., Strauch, A., Tottrup, C., Truckenbrodt, J., & Geijzendorffer, I. R. (2018). Mapping Mediterranean wetlands with remote sensing: A good-looking map is not always a good map. *Advances in Ecological Research, 58*, 243–277. doi:10.1016/bs.aecr.2017.12.002

Ramsar Convention on Wetlands. (2018). *Global Wetland Outlook: State of the World's Wetlands and their Services to People. Ramsar Convention Secretariat, Gland, Switzerland Ramsar Convention Secretariat. (2016). The fourth strategic plan 2016-2024. Ramsar handbooks for wise use of wetlands*. Ramsar Convention Secretariat.

Scaramelli, C. (2020). The Lost Wetlands of Turkey. *Middle East Report (New York, N.Y.)*, *296*, 14.

Şekercioğlu, Ç. H., Anderson, S., Akçay, E., Bilgin, R., Can, Ö. E., Semiz, G., & Dalfes, H. N. (2011). Turkey's globally important biodiversity in crisis. *Biological Conservation*, *144*(12), 2752–2769. doi:10.1016/j.biocon.2011.06.025

Sıkı, M. (2020). *Gediz deltasi'ndaki İzmir kuş cennetini yaşatma mücadelesi (1982-2018)* [The struggle to keep the Izmir bird paradise alive in the Gediz delta (1982-2018).]. Kanguru Yayınları.

Stein, A., Gerstner, K., & Kreft, H. (2014). Environmental heterogeneity as a universal driver of species richness across taxa, biomes and spatial scales. *Ecology Letters*, *17*(7), 866–880. doi:10.1111/ele.12277 PMID:24751205

TUIK TSI. (2020). Address Based Population Registration System Results, 2019. [Data set]. TurkStat.

Weise, K., Höfer, R., Franke, J., Guelmami, A., Simonson, W., Muro, J., O'Connor, B., Strauch, A., Flink, S., Eberle, J., Mino, E., Thulin, S., Philipson, P., van Valkengoed, E., Truckenbrodt, J., Zander, F., Sánchez, A., Schröder, C., Thonfeld, F., & Hilarides, L. (2020). Wetland extent tools for SDG 6.6. 1 reporting from the Satellite-based Wetland Observation Service (SWOS). *Remote Sensing of Environment*, *247*, 111892. doi:10.1016/j.rse.2020.111892

Zedler, J. B. B., & Kercher, S. (2005). Wetland Resources: Status, Trends. *Ecosystem Services, and Restorability. Annual Review of Environment and Resources*, *30*(1), 39–74. doi:10.1146/annurev.energy.30.050504.144248

Section 2
Urban Wetlands

Chapter 3
Integrated Urban Freshwater Landscape at Risk:
Ecology, Governance, and Sustainability of Urban Wetlands in Colombo, Sri Lanka

M. D. K. L. Gunathilaka
https://orcid.org/0000-0001-7406-2800
University of Colombo, Sri Lanka & Yunnan University, China

W. T. S. Harshana
Urban Development Authority, Sri Lanka

ABSTRACT

Freshwater wetlands are more threatened than any other ecosystem in the world. The current trend of freshwater wetland crisis arises a question 'whether it is conversion or conservation of wetlands' which is absolutely a dilemma. Therefore, this study aims critical evaluation at wetland governance and sustainability of urban wetlands in Colombo. The results reveal that the Colombo wetlands have lost their territory from a moderate to a severe degree. The wetland loss is higher in Kaduwela, Kolonnawa, and Kesbewa areas. Recently the causes of Colombo wetland loss has moved with a new frame. Land grabbing and green grabbing are the cycles that transforming the geostrategically important urban wetlands into hybrid ecosystems. Due to this, the frequency of flash flood events and inundation time has increased than ever. It is important to ensure urban freshwater wetland sustainability and surveillance following sustainable development goals and it is time to have freshwater ethics.

DOI: 10.4018/978-1-7998-9289-2.ch003

INTRODUCTION

An ecosystem has always been made up of biotic and abiotic matter, including all the physical, chemical, and biological processes that operate on Earth. Ecosystems are dynamic, whether they are small or large. That is, damage to a part of this system directly or indirectly affects the entire ecosystem. Among the ecosystems, freshwater ecosystems are more sensitive and their relationship dates back to early history. The world's oldest civilization is based on where water resources are available. That is, the formation of early civilizations like Mesopotamian Civilization also known as the Cradle of Civilizations along the Euphrates and Tigris valleys also known as the Cradle of Civilizations or the Fertile Crescent between 4000 and 3500 BC; another example is the emergence of Egyptian civilization around the Nile Valley around 3200 BC (Gokce, 2018) Besides, the Hwanhuou or Yellow-river Civilization in China and Indus-Valley in Northwest India are great examples of civilizations that flourished along a river. Thus, it has now been proven that there are some key factors behind the formation of settlements near water bodies. Among them, agriculture is the prime. Adequate water, as well as land and alluvial soil, are the essential elements of agriculture. Accordingly, this need was met by water-based river valley lands. These ancient civilizations provide evidence that paddy cultivation was the mainstay of agriculture in these lands. The lands used for this purpose were river valleys on both sides of submerged rivers, most of which were later known as wetlands. That is, the wetland is the first ecosystem to be affected by mankind on Earth.

Globally, wetlands are distributed in every biome with an exception of Antarctica. It is approximately 6% of Earth's land surface covered by wetlands (Reddy & DeLaune, 2008). Being complex ecosystem wetland systems are driven by various physical, chemical and biological processes. In the beginning, scientists used to explain wetlands by their life form, hydrology, morphology, chemistry, vegetation or microbiology (Reddy & DeLaune, 2008). It was like that fabled group of blind men trying to describe an elephant that was completely a mess. Extensive research specializations in wetland science in the 20th century paved the way to further experiments of critical features of the global landscape because of their unique role in regulating global biogeochemical cycles, value and function of the wetlands. Wetlands are located in areas with low elevation and higher water tables (Reddy & DeLaune, 2008) including those of marshes, swamps, bogs and similar areas under a comprehensive definition. Different classification systems of wetlands have been developed to function for various purposes. United States Fish and Wildlife Service in 1979 presented a classification of wetlands and deepwater habitats of the United States which is a hierarchical system based on marine, estuarine, riverine, lacustrine and palustrine (Matthews, 2013). Similarly, the Ramsar Convention made wetland classification based on the US classification which was modified by adding it to

Figure 1. Global distribution of wetlands
Source: Finlayson et al., 2016

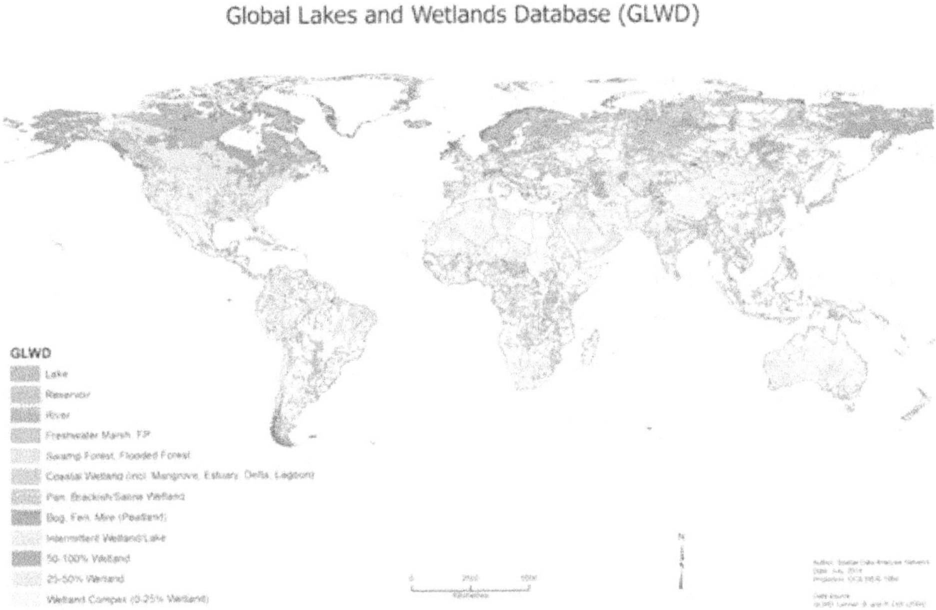

human-created or cultural wetlands. Other classifications are based on the structure or function or management goals (Crandell, 2021). Among the most used definition of wetlands is the Ramsar definition in 1971.

Areas of marsh, fen, peatland or water, whether natural or artificial, permanent or temporary with water that is static or flowing, fresh, brackish or salt, including areas of marine water the depth of which at low tide does not exceed six metres (Kotagama & Bambaradeniya, 2006).

According to the above definitions, a wetland is a lowland area relative to the external physical environment and is also an area of greater biodiversity with a waterlogged soil zone most of the year. The distribution of wetlands are varied in size, functionality and geography (Figure 1) (Finlayson et al., 2016).

During the 1980's, attention emerged for conservation of Asian wetlands. As a result, two publications were released in 1989; 'Overview of Asian wetlands' and 'Directory of Asian wetlands'. The directory listed 947 Asian wetlands (73 ha in million) of high international importance. The total area of Asian wetlands was 120 million hectares. 80% of them occurred within seven countries; China, India, Papua New Guinea, Bangladesh, Burma and Vietnam (Zon, 2003). In terms of wetland distribution in India, 67429 wetlands can be identified and their total land area is 4.1 million hectares (MoEF, 2020) (Ministry of Environment & Forest (MoEF). Of these, 2175 are natural wetlands and 65254 are man-made. Considering the distribution of wetlands in Bangladesh, it appears that the wetlands represent an area of 5765,000 hectares out of the total land area (Sulthana & Thompson, 2010). In the case of Pakistan, statistics show that 1888 wetland ecosystems identified under the Ramsar Convention cover an area of 18553 million square meters. Inland wetlands in Pakistan extends over 7800000ha (Scott, 1989). Asian wetlands yield a wide range of goods and services having high economic value and rich biodiversity.

The Asian wetland directory listed 41 wetlands (274,000ha) of internationally important in Sri Lanka (Zon, 2003). Wetlands in Sri Lanka can be matched to the Ramsar definition broadly divided into three categories. They are a) inland natural freshwater lands *i.e.* rivers, streams, villus, marsh, and swamp forest, b) marine and saltwater wetlands *i.e.* lagoons, estuaries, mangroves, sea-grass beds, and coral reefs, c) man-made wetlands *i.e.* tanks, reservoirs, rice field, and salterns (Kotagama & Bambaradeniya, 2006). This study paved the way to discuss urban freshwater landscape which can be broadly listed as inland freshwater wetlands. The freshwater wetlands in Sri Lanka that fit into the Ramsar definition can be divided into two categories; a) inland natural freshwater wetlands i.e. rivers, streams, waterfalls, marshes, swamp forests, freshwater springs, montane bogs/swamps, and villus, b) man-made freshwater wetlands i.e. tanks, rice fields, irrigation canals, aquaculture ponds, tanks, reservoirs (Bambaradeniya, 2003). In this respect, the main urban area in Sri Lanka is distinguished for its integrated freshwater wetlands. An integrated freshwater landscape consists of three main ecosystems; stream, lake, and marshes. All of these wetlands comprise a mixer of soils, water, plants and animals. The interaction between these elements allows wetlands to perform a wide scale of functions that are beneficial to all living beings, humankind in particular.

The direct and indirect role played by these wetlands in maintaining the balance of the entire globe is magnificent. Wetlands are one of the most productive ecosystems on earth; their functions and values have only been distributed among the community in the past fifteen years. According to the Constanza et al. (1997) economic analysis of various ecosystems showed that freshwater wetlands and estuaries are more valuable than other ecosystems of the biosphere. The importance of wetlands are varied; acting as a reservoir of excess water; as sponges to absorb excess water, contributing to the maintenance of biodiversity, climate regulation role played as sinkers of greenhouses gases and minimizing the scale of global warming, agricultural activities, economical value for pharmaceutical and cosmetic industry, fishery, recreation and tourism, as a means of transportation, hydro-electricity and aesthetic values (Bambaradeniya, 2003; Kotagama & Bambaradeniya, 2006). Wetland plays a unique role in controlling the risk of flooding, which is a natural disaster in an area. This excess water, which is stored in wetlands, functions as a water reservoir, meeting the water needs of all living organisms in the area. In addition, the unique plant species present in the wetlands filter the stagnant water in the wetlands and store the filtered water underground. As it is saturated with water when it is wet, it meets the water requirement which is one of the most important elements required by living beings. For these reasons, a large number of species inhabit wetlands including birds; terrestrial or waterbirds, resident or migratory; mammals, fish, amphibians, and reptiles as well as a vast range of microbes. These include both the flora and fauna and globally as a whole, several plants and animal species carry out their biological activities based on these wetland sources. Pantanal is the world's largest freshwater wetland. It is also home to 3,500 plant species, 1,000 species of birds, 400 species of fish, 300 species of mammals, 480 species of reptiles and insects based on the Pantanal wetland north of the world's largest wetland, the Paraguay River, and its tributaries. It has been explored that about 900 faunas live in this single wetland unit alone (DWC, 2016). Displaying richness and uniqueness in biodiversity; *i.e.,* over 2000known species of freshwater fishes, riverine wetlands; Nile, Niger, Zaire and Zambezi surrounded several freshwater wetlands in Africa. The Zaire riverine wetland is home to 560 endemic fish species (Kabii, n.d.). Lake Ichkeul in Tunisia is a Ramsar wetland is a home to nearly a million migrating birds yet to be investigated. Tunisia alone designated 41 Ramsar wetlands indicating its rich biodiversity and ecological importance. Approximately 250,000km^2 out of 650,000km^2 of wetlands in China have been reserved. Thus, wetlands have known for their unique biodiversity all over the world. Wetland plants are at the base of the food chain of a wetland ecosystem are of four types; a) emergent, b) submerged, c) floating-leaved and d) floating. Hence, wetlands are open to the resident as well as migratory and invasive species which in turn affect the surveillance of the ecosystem. Once a wetland is rich in biodiversity, it provides a greater variance of ecosystem services. In Sri Lanka, 30% of inland vertebrate species are ecologically dependent on inland wetlands. For example, 60 species of primary freshwater fishes, of which 25 are endemic. Among the inland invertebrates, freshwater wetlands are home to more than 40 endemic species of crabs. About a third of bird species are dependent on freshwater wetlands (Bambaradeniya, 2003). Food chains and networks are created based on the interactions between these plant and animal species, and these food networks and chains make the interaction process more active in the wetland ecosystem. Through it, even biodiversity that has not yet been explored and researched can exist in these systems. Accordingly, it can be said that many important biological processes associated with these wetland ecosystems are maintained with biodiversity. That is to say, these wetlands play a vital role in maintaining biodiversity.

Another important function of a wetland ecosystem is to absorb and store most of the greenhouse gases that contribute to global warming. A recent study carried out by, Dayathilaka et al., (2020) found that Above Ground Biomass carbon stocks in the wetland reserve 19803 ± 1566 tCO_2 eq. and in the Thalawathugoda wetland area 4180 ± 729 tCO_2. This suggests that these wetland complexes play an important role as carbon sinkers. That is, how these wetland ecosystems function as an air filter controlling global warming is critical to atmospheric balance.

Most of the global economic activity takes place in wetlands. Subsistence or market-based agricultural activities are highly dependent on wetlands, rice farming in particular. This agricultural pattern was combined with the cultural identity and cultural importance of wetlands thus highlighted since past. In Sri Lanka, the cradle of historic civilization was those inland freshwater wetlands; floodplains and rivers and then the ancient cascade of tanks. Other than agriculture, wetlands supply firewood for domestic purposes, reeds for handicrafts and local designs *i.e.* *Cerbera mangas* and Pandanus leaves used in Sri Lanka, medicinal herbs *i.e.* Sri Lanka, *Bacopamoneri* and *Hygrophylla shalli* are used in indigenous medicine and response for the demand of protein consumption of a country. Hydroelectric power generation *i.e.* Victoria, Rantamben and Radenigala in Sri Lanka reservoirs and promotion of tourism. Hydroelectric power promotion of tourism especially in main tourist destinations in the world *i.e.* two million hectares of Kakadu National Park in Australia and the Tsomoriri Wetlands in Kashmir (Wetland Tourism, 2012) are trends within the past 30 years. In this regard, wetland ecosystem services have both direct and indirect benefits gained by people through the exploitation of wetland resources.

Notwithstanding the importance, wetlands in the world are at a greater risk. Approximately 35% of the world's wetlands were lost for 45 years from 1970 and the loss rate is accelerating annually since 2000. Up to 40% of the world's species live and breed in wetlands. Now more than 25% of all wetland flora and fauna are at risk of extinction (UNFCC, 2018). Wetlands are disappearing three times faster than forests (UNFCC, 2018). Figure 2 shows the decreasing trend of wetlands by continent. The wetland extent decreasing trend is faster in Asia and Europe. And this trend is particularly reported from freshwater wetlands. The Freshwater of the world is collectively experiencing accelerating rates of quantitative and qualitative degradation (Wetzel, 2001). The recent loss of freshwater wetlands worldwide is valued at $2.7 trillion per year (Postel, 2019). This loss directly harms the wetland functions including mitigation of floods, protecting from the storm, retention of sediments, purification of water and removal of toxic compounds, carbon sink, and breeding habitats for edible fish species. Therefore the signs of immense threats on freshwater wetlands in Sri Lanka also have been visible. Habitat deterioration and degradation have been the topmost threat gradually increasing the degree of

Figure 2. Wetland loss in the world
Source: Ramsar fact sheet 3, n.d.

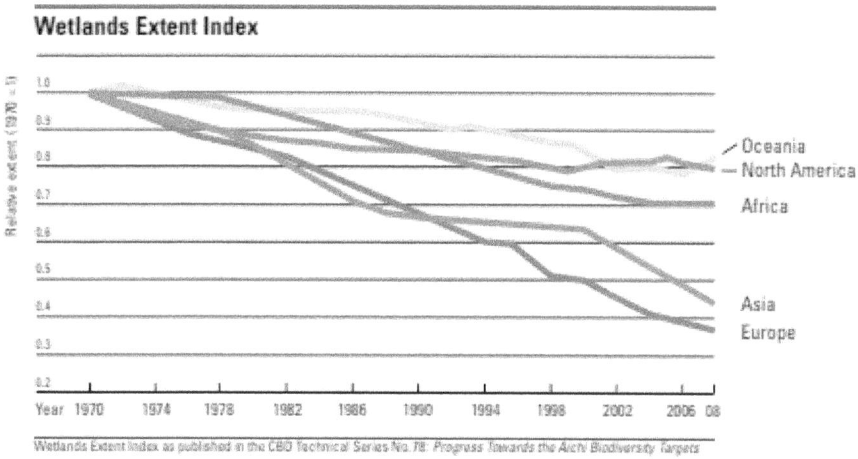

freshwater wetland degradation. Direct loss and exploitation of species due to various anthropogenic pressures is the second most threat while the spread of invasive alien species has become a silent but aggravated threat on freshwater wetlands in Sri Lanka. Each threat is comprised of several causal factors and their impact on freshwater wetlands are enormous. Therefore it is no doubt the freshwater wetlands in Sri Lanka is at a greater risk. Especially freshwater wetlands in urban areas. Therefore this study identified the timely importance of discussing the Colombo freshwater wetlands and their ecology and governance and sustainability. This study aims to quest whether the current scenario of urban freshwater wetland governance truly minimizes the anthropogenic climate change impacts. Targeting towards the aim, the study pays attention to a) Colombo wetlands and their service and their ecological importance; b) Current status along with spatio-temporal wetland changes using GIS and Remote Sensing techniques; and c) the governance of the urban wetlands in Sri Lanka under climate change. Hence the study followed mix-method to fulfil the aim of the study.

COLOMBO URBAN FRESHWATER WETLANDS

Considering the ecological status, Colombo is a city surrounded by a large and interconnected system of natural wetlands, originated due to the flow of the River Kelani through the Colombo district, providing precious flood control service. Since the pre-colonial era wetlands have done a great service for the city of Colombo and its

Figure 3. Colombo wetlands
Source: Author, 2020

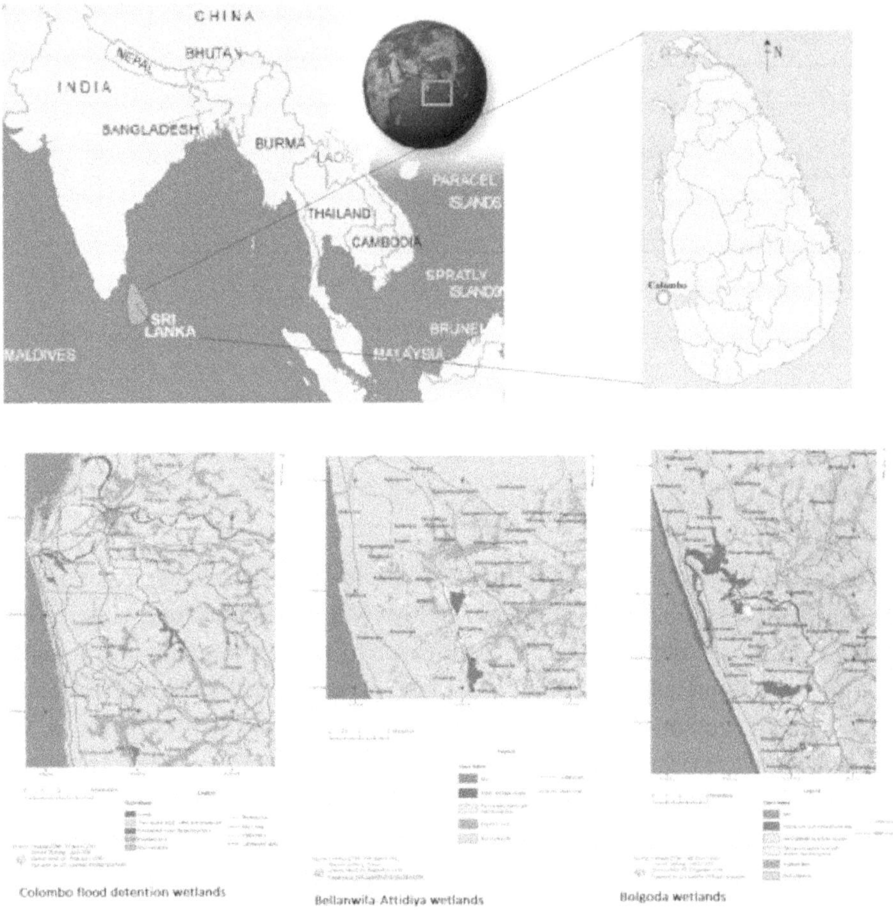

suburbs protecting from the flood absorbing access water, in particular. Wetlands are low lying marshy areas. The wetlands in Sri Lanka which come under the Convention mainly divided into three as 1) inland natural freshwater wetlands, 2) marine and saltwater wetlands and 3) man-made wetlands (Kotagama and Bambaradeniya, 2006). Colombo belongs to the first type in the wetland category which possesses a large extent of riverine and palustrine wetlands. Such as deltas, permanent rivers, floodplains, permanent freshwater lakes, permanent freshwater marshes, peatlands, shrub-dominated marshes, tree-dominated marshes and springs.

Colombo wetland ecosystems cover an area of about 20km^2 in the Colombo District and are comprised of Diyawannawa, Beddagana, Thalawathugoda, Kotte, Heen Ela, Madinnagoda, Mulleriyawa, Kolonnawa, Thalangama and Ambatale

Wetland Complexes (Figure3). Apart from them, peripheral wetlands located in Maharagama, Dehiwala, Kesbewa areas are also significant. According to the Environmental Foundation Limited, it has been observed that these wetlands carry 39% of the flood disaster out of the city during heavy rains in the Colombo District. In addition, about 60% of the population living in the Colombo Metro Region (CMR) make a living from these wetlands. Providing traditional medicine, cooling 50% of the city of Colombo, storing about 1.43 million metric tons of carbon released by the CMR, as well as inhabiting more than 250 endangered and endemic plant species. The wetlands in the city of Colombo play an important role in providing habitat for about 280 species of animals (MCUDP, 2016). Accordingly, the contribution of these wetlands is very high in controlling the flood situation which is a serious problem in the Colombo District (Ranasinghe, 2018). Diywanna Wetland Complex, Dehiwala-Attidiya Wetland Complex and Kolonnawa Marsh are the three main integrated wetlands in the Colombo District mentioned above. All these wetlands are located in the lower river Kelani basin.

Kolonnawa Marsh can be identified as a freshwater marsh system located in the Colombo District and this system is based on the Kolonnawa Wetlands and Heen Canal (Marawila & Tibbotuwawa, 2010). It can be seen that a significant portion of this land is currently being used for physical reclamation and paddy cultivation, leading to the gradual destruction of the ecosystem (Central Environmental Authority/ CEA) (IUCN. & CEA., 2006). The Diyawanna wetland is considered to be one of the three major wetland ecosystems in the district, namely Kolonnawa Marsh, Kotte Marsh and Heen Ela Wetlands, which represent the three main flood control zones associated with the Diyawanna wetland. The total system is about 400 hectares, of which about 200 hectares are represented by the Diyawanna Wetland Complex which surrounds Parliament. This wetland system, with its high flora and fauna diversity, is based in the Colombo middle-class area (Marawila & Tibbotuwawa, 2010). The Bellanwila Attidiya Wetland, which is based on the upper catchment area of the Bolgoda River, is another unique ecosystem representing the Colombo District. The wetland system covers an area of 372 hectares (IUCN. & CEA., 2006) and is located at an altitude of 0.6m above sea level. The average temperature is approximately 27°C and the annual rainfall is about 2500mm (Henkanaththegedara, 2005). Rice was cultivated until about 1978 and this ecosystem, which is rich in biodiversity, earns a large number of people in the neighbourhood (IUCN. & CEA., 2006). The site was declared a sanctuary by the Flora and Fauna Protection Ordinance in 1990 and was conserved as a wetland by the CEA in 1992 (IUCN. & CEA., 2006).

Kelani River Basin is also home to over 25% of the Sri Lankan population (IUCN, 2018). Among them, the highest agglomeration of the population can be seen in the lower basin and it is very closer to the river mouth. The transformation of eco-social landscapes often results in novel and hybrid ecosystems that differ from the original ecosystems both in an ecological and social setting. Watershed urbanization that has been taking place, hydrological modification and the changes in ecosystem use are the main causes of Colombo wetland transformation (Hettiarachchi, 2014). According to the flood occurrence data, annual floods during May can be expected more often in Colombo. But due to the unawareness of the natural phenomenon or the lack of knowledge on the general geographical phenomenon, different types of anthropogenic activities have recently gathered on the flood plain of Kelani River. As a result of this Colombo, flooding has become a more devastating issue (Dammalage & Jayasinghe, 2019). But compared to the present scenario of the Colombo region, in past, there had been more wetlands to absorb the excess water when the Kelani River had overflown. A study carried out by Missaka et al., (2014) found that the marshy environment of the Colombo wetland is rapidly transforming into a habitat with shrubs and small trees replacing native grass. Watershed transformation to urban and suburban land uses and land cover intensified the degree of frequent flooding than before (Dammalage & Jayasinghe, 2019). Typical peat soil in the Colombo marsh has also changed into semi-mineral soil. Both changes result in a significant reduction in the water holding capacity of the wetland, thus increasing the flood frequency. The contribution of wetlands to mitigate the impact of floods have been reported in other parts of the world where the loss of wetlands and increased appearance of new flood paths has been highlighted. With the decline of these wetlands, the absorption process has disturbed and there was no place for water to infiltrate. Urbanization based land-use change stimulus the local hydrometeorological processes result in altered the urban micro-climate (Pathirana et al., 2014).

ECOLOGY OF PIONEERING WETLAND COMPLEXES

These integrated wetlands are ecologically and hydrologically important. Based on the location of the centre of the Bellanwila Attidiya freshwater wetland, which is developing rapidly, plays an important role in controlling the flood situation in the area. The size of the catchment area around the wetlands is 17.6 km^2. This catchment area is divided into seven sub-feeder areas (CEA, 1993). The Bolgoda Basin and the Verus River Basin are the main sources of water. The groundwater level here is very high when it comes to groundwater. It is located a few centimetres below. This varies between dry and rainy seasons, but there are almost no extreme cases. Surface water supply is mainly through Bolgoda canal, Oil canal and Katu canal (CEA, 1993). Salinity is a major factor in the quality of marsh water in the area. Accordingly, the salinity of the Bellanwila-Attidiya marshland was tested during the Bellandawila-Attidiya marsh research wetland site report and conservation management plan conducted by the CEA of Sri Lanka and the Netherlands in 1993. The salinity of the main water basin, the Bolgoda Basin, has been recorded at a very low level. It has an average value of around 200 - 250 ppm. However, the salinity around the Bolgoda Reservoir is slightly higher and it is 380ppm. Further, the lower zone of this wetland is slightly more saline due to the inflow of ocean water through the Panadura River. Overall, the Bellanwila-Attidiya wetland region has both suitable and unfit water bodies suitable for human consumption. The results of the research also show that this water can be used for cultivation and it is suitable for cultivation after treating the water in the Oil canal (CEA, 1993).

As far as the Kolonnawa wetland is concerned, it is spread over 15 Grama Niladhari Divisions (GND) and the swamp is spread over only ten GNDs. The land is fed by several water basins and research on wetlands has revealed that the catchment area associated with this wetland in 2010 was 144.56 hectares. This is a gradual decrease in the land area since 1992 (Samarasinghe & Dayawansa, 2013). The count of faecal coliform bacteria has reached almost 23000 per 100ml. This is almost 23 times the standard value fixed by the WHO. Toilet waste also has increased phosphate levels in the water to almost nine times the ambient water quality standard (0.1mg per litre) (Sushmita, 2012). Accordingly, it can be said that the water is polluted. In discussing the Diyawanna wetland complex, it can be stated that this swamp is fed based on Diyawanna Oya. The surface water quality was tested by the Urban Development Authority (UDA) under the consultancy of the University of Moratuwa in 2011 and the behaviour of the water parameters during its period from 2004 to 2009 (Table1) (UoM, 2011). Furthermore, groundwater levels are estimated to be between 0.4-0.5m in the shallows. Research has also shown that it is 1.5-3.0 m above the surface (MCUDP, 2011). According to these results, the water associated with the Diyawanna Oya is also polluted to somewhat extent. It can also be pointed out that the main reason for this is the influence of human processes.

Considering the ecological features associated with this wetland trinity, it is clear that there are three main wetlands in the Colombo District that have high biodiversity. Given the Bellanwila-Attidiya wetland, many species of flora and fauna have been identified and listed at present. Accordingly, fish species are important among the identified species. Thirty-two species of fish were identified by Pethiyagoda in 1991 which is cited by Henkanaththegedara in 2005, including four endemic species, seven introduced and one new species. Among the introduced species are the most common fish species: *Oreochromis mossambicus*, *Oreochromis niloticus* and *Hypostomus plecostomus*. *Puntius vittatus*, *Horadandiya atukoralai* and *Aplocheilus parvus* also grow rapidly in the shallow waters (Henkanaththegedara, 2005).Several endangered fish species can be identified based on this wetland. Namely, *Horadandiya atukoralai*, *Puntius bimaculatus*, *Aplocheilus dayi* and *Channa orientails* (CEA, 1993).

Also, Bolgoda rises during seasonal tides and normal tides, *Megalops cyprinides*, *Tuarine* species, *Amblypharyngodon melettimus*, *P.chola*, *P.amphibions*

Table 1. Water quality in Diyawannawa wetland complex

Parameters	Water Quality (during 2004-2009)
pH	5.8-6.9
Conductivity (ds/m)	0.1-0.3
Turbidity (NTU)	8-15
Ammonia (mg/l)	0.24-0.56
Nitrates (mg/l)	0.3-0.4
Phosphates (mg/l)	0.3-0.4
DO(mg/l)	3.2-5.5
BOD (mg/l)	6-10
COD (mg/l)	17-33

Source: UoM, 2011

(Henkanaththegedara, 2005). In the discussion of bird species, Vardhana identified 153 species of birds representing a total of 48 families today, including 76 species in 1980, 89 species in 1987 and 152 species in 1993. Meanwhile, *Phalacrocorax fuscicollis*, *Rallus striatus*, *Porzana fusca* are rare bird species that have been identified. Several other bird species have been recognized, including the *Pelecanus philippensis*, *Ixobrychus flavicollis*, *Mycteria leucocephala*, and *Anastomus oscitans* along with 30 species of snakes are listed under 13 families (CEA, 1993). Among them are endemic species, *Cylindrophis macutatus*, *Lycodon osmanhilli*, *Oligodon sublineatus*, and *Xenochrophis asperrimus,* and endangered *Python molurus* lives inhabiting special wetlands. *Crocodilus palustris* and Otters are also found in these wetlands (CEA, 1993). When discussing wetland flora, floating plants including, *Eichornia crassipes*, *Salvinia molesta* and Monochoria can be introduced. Submerged rooting plants include *Cabomba* Sp., *Hydrilla verticillata* and *Nymphea lotus* Colu). Further, *Anona glabra*, *Jussiea peruvin*, *Colocasia* Sp., *Commelina diffusa*, *Jussiea repens*, *Jussiea suffruticosa*, *Cebera manghas*, *Nymphea lotus*, *Cobomba* Sp., plant species can be identified in the Bellanwila-Attidiya wetland. Herath (1992) observed 46 species of vascular plants, all of which belong to 34 families (CEA, 1993). When discussing wetland soils, this region is covered with an ancient crystalline layer, and according to Cooray (1984) Charnokitic, Hornblende, Biotite gneisses and Granite are found rocks (CEA, 1993).

Kolonnawa Wetland Complex has become a habitat for many species of plants and animals. Meanwhile, 85 species of plants have been identified in terms of wetland flora, of which 6 are endemic species. Furthermore, one species is critically endangered and one species is threatened. Also, two other species are endangered (MCUDP, 2016). Mainly *Aganope heptaphylla*, *Hypolytrum nemorum*, *Cinnamomum zeylanicm*, *Areca concinna* and *Quassia indica* are distributed. Several invasive plant species have also been identified in the wetlands, namely, *Annona glabra*, *Eichhornia crassipe*s and *Salvinia molesta* (MCUDP, 2016). In terms of population, snails identified are not endemic or invasive, and one species is exotic. Eight species of snails have been identified, one of which is exotic. Observations have identified 20 species of dragonflies, one of which is endangered and the other endangered. In addition, 25 species of butterflies have been identified in connection with the Kolonnawa Wetland Complex, one of which appears to be endangered. In considering freshwater fishes, 14 species of fish have been identified, two of which are endemic. Three species have been identified among the snake species. 62 species of birds have been observed in the wetlands, of which two are endemic and three are exotic. Four species of mammals have been found, one of which is endemic. The *Onychargia atrocyana* of the family Coenagrionidae is one of the most endangered species of butterflies, and the *Ideopsis similis* of the Nymphalidae family is one of the most endangered species. Among the endangered species is *Semnopithecus vetulus*, a member of the Cercopithecidae family of mammals (MCUDP, 2015). Considering the soil characteristics associated with wetlands, soil moisture is 45-60%, pH is in the range of 3.5 to 4.7, electrical conductivity is between 1100-2100 s/cm and sand content in the soil is between 20-25%. Also, the colour of the soil is very dark brown, with a moderate level of organic matter content of 40-60% (Ranasinghe, 2019).

The Diyawanna Wetland Complex is also an ecosystem rich in flora and fauna. Based on the wetland, 89 plant species have been identified, of which 13 are exotic species and three are invasive species. One endemic species has been identified among those species. Among these are two endangered species. Two other species have also been identified as endangered. Seven species of invasive species are *Annona glabra*, *Alstonica macrophylla*, *Dillenia suffruticosa*, *Panicum maximum*, *Pennisetum polystachion*, *Eichornia crassipes* and *Salvinia molesta* (MCUDP, 2015). Many species of snails can be identified in wetlands when discussing animal species. Six species of snails have been identified. An endangered species of snail has also been identified. Meanwhile, seven species of water snails inhabit the wetlands. Fifteen species of dragonflies inhabit the wetlands, one of which is endemic. It has also been discovered that 17 species of butterflies have created habitats based on wetlands. Fifteen species of fish have been observed in the discussion of fish, two of which are endemic. Six species are endemic and one species is endangered. Five species of amphibians inhabit this wetland, one of which is endemic. Seven species of snakes have been reported in the wetlands, one of which is threatened with extinction. 43 species of birds are recorded based on wetlands, one of which is endemic. Two exotic species have also been identified. Among these species are from the migratory party *Charadrius placidu*, *Pluvialis fulva*, *Pluvialis squatarola*, *Calidris minuta*, *Gallinago stenura*, *Ixobrychus sinerisis*, *Motacilla flava*, *Actitis hypoleucos*, *Tringa stagnatilis*, as well as resident bird species *Anastomus oscitans*, *Dendrocygna jvanica*, *Bubulcus ibis*, *Phalacrocorax fuscicollis*, *Porphyrio porphyrio*, *Fulica atra*, *Ardeola grayil*, and *Prinia socialis* are some of the bird species observed (MCUDP, 2015). *Onychargia atrocyana, Scabrina brounoe, Aplocheilus dayi, Crocodylus porosus, Semnopithecus vetulus* and *Prionailurus viverrinus* can be identified as some of the most endangered species in this wetland as a whole (MCUDP, 2015). In terms of soil characteristics, the researchers point out that the soil characteristics of this melting wetland are very similar to those found in the Kolonnawa wetland complex (Ranasinghe, Piyadasa, & Mataraarachchi, 2019).

CURRENT STATUS OF THE INTEGRATED FRESHWATER LANDSCAPE

The rapidly receding boundaries of population agglomeration in Colombo has given rise to many environmental problems demanding urgent attention. During the 1996s' the population in the built-up areas of Colombo is 160 persons per hectare (Kawamura et al., 1996). When considering the built-up area in the past and present a heavy difference can be seen. If take a look at CMR only, the built-up area in 1992 was 11165ha which had increased to 19393ha in 2001 and 35876ha in 2016. This depicts that the built-up area has increased by approximately 221 per cent in the past 22 years from 1992 to 2014 at a rate of 1123ha per year (Subasinghe et al. 2016). Since the establishment of the port in Pettah in 1833 a large volume of non-residential activities such as stores, warehouses, container yards, electricity sub-sections and oil storage tank complex have been accommodated around suburbs, in particular, Kolonnawa, Kaduwela, Maharagam, Kesbewa, Homagama areas, demonstrating urban expansion (Panditaratna, 1965). In the beginning, the elite lived in the core of the city. However, gradually the elite group was replaced with low-income urban poor who commuted as labours. In 1974 there were 16251shanties with 19953 families living in them with a total population of 90645. There were 20507 houses in slum tenement gardens with 24623 families and a population of 123115. In addition, there were 8207 slum tenement houses with 9840 families and a total of 48907 people living in them. The total number of slums and shanty dwellings was 44965 and the number of people living there was 271667. To satisfy the special requirements of these growing populations, the land use and the land cover of ancient Colombo has been altered into human settlements and shanties. Accordingly, Kolonnawa, Colombo and Kaduwela along with other suburbs in Colombo, has undergone rapid land-use alteration.

The magnitude of land-use alteration was identified during the 2000s. Before 2000, the urban land change accounted for 3.5 per cent of the entire landscape, while during the 2000s it was accounted for 7.02 per cent. After the mid-2000s, the decrease in the productivity of agriculture in the dry zone of the country prompted the government to push rural to urban migration as a poverty-reducing initiative (Hewavitharana, 2004). This influenced the young generation and motivated them to migrate to settle in Colombo, resulting in urban expansion. The ultimate effect of this urban expansion is land fragmentation. This has a direct effect on landfills, particular in, wetlands and marshy areas. Thus, people expanded towards Kolonnawa then Battaramulla to Kaduwela suburbs and gradually towards the inland are highly significant. The socio-economic characteristics of the study area emphasized the grey view of the increasing trend of population and housing while reducing the wetland area. Samarasinghe and Dayawansa (2013) have highlighted 41 per cent of the surveyed sample are lack land ownership and are considered illegal encroachers. Among the illegal encroachers, Obesekarapura and Welikada West Grama Niladari (GN) divisions majored in the Kolonnawa Divisional Secretariat Division (DSD) area. Among the factors that affect for degradation of wetlands in this area, invasiveness, changes to the water regime are two main factors (Shine & De. Klemm, 1999). Since wastewater and stormwater drainage has been a greater issue in Kolonnawa these wetlands become vulnerable in various ways. Marsh areas are the most utilized sites for dumping solid waste and liquid waste. Elevated nutrient levels created by solid and liquid wastes, excreta and animal wastes influence on excessive growth of invasive plant species and algae where interrupt the natural infiltration and flow rate of water in the marsh. Commuters in use Kolonnawa road network also used to throw away their solid wastes in this area (Samarasinhe & Dayawansa, 2013).

The interest in environmental protection has decreased among the community. As revealed by Samarasinghe and Dayawansa (2013) a large scale of landfilling and reclamation has taken place in Kalapaluwawa, Kotuwegoda and Subhuthipura GNDs where there is no organization to raise their voice to protect these wetlands. Further Samarasinghe and Dayawansa (2013) carried out a wetland risk analysis which revealed 90 per cent of wetlands covered with *Annona glabra* and *Eichhornia crassipes*. Also, firewood collection and cultivation within the marsh had been identified as another biggest silent threat. The risky land use areas were Obesekarapura, Welikde West and Salamulla GNDs where the physical change was at high risk. This implies the high number of landfilling activities that took place during the 1992 and 2010 periods along with the illegal encroachments. Gajabapura and Welikade Esat showed a moderate risk. As mentioned earlier unmanaged disposal of solid and liquid waste has made the marsh in Obesekarapura and Kotuwegoda at high risk. The biological products were also identified as high risk in the Obesekarapura wetland area. Samarasinghe and Dayawansa (2013) accordingly concluded waste disposal in both solid and liquid types, illegal drainage pipes of human excreta discharge, reclamation and landfilling, high population density, illegal encroachments and

Figure 4. How media report Colombo wetlands
Source: Google.com, 2021

dispersion of invasive species as noteworthy threats in wetland degradation in the Kolonnawa area. Not only have the researchers the media also reported the risk of Colombo wetlands (Figure4).

Another example is the Bellanwila-Attidiya marshes. Spatial analysis performed by the government (2018) identified that the number of buildings in the protected area increased from 5158 in 2004 to 7524 in 2017. It indicates an annual increase of 2.95%. Landfilling for development, increase in population, and socio-economic activity resulted in further contamination of the marshes. The wastewater contamination is severe, threatening the ecosystem as well as ecosystem services (GoSL., 2018). The clearance of wetland areas, illegal reclamation, dumbing garbage, and encroachments are quite common in Diyawanna wetlands as well (Marawila & Tibbotuwawa, 2010). A significant decline in the water quality of Diyawanna lake is visible. The water is discoloured and green sediments of algal blooms are there. Wastewater and other effluents are released to the lake via illegal underground pipelines. Figure 5 further proves the gradual decrease of freshwater wetlands within the Colombo freshwater landscape. The NDVI classification further proves the spatio-temporal variations of wetlands in the area (Figure 6). The most of land reclamation and construction that has taken place around Colombo wetlands are for recreational and commercial purposes. This process is extensively discussed under the wetland governance.

Figure 5. Colombo wetland changes by division
Source: Author, 2019 (unpublished)

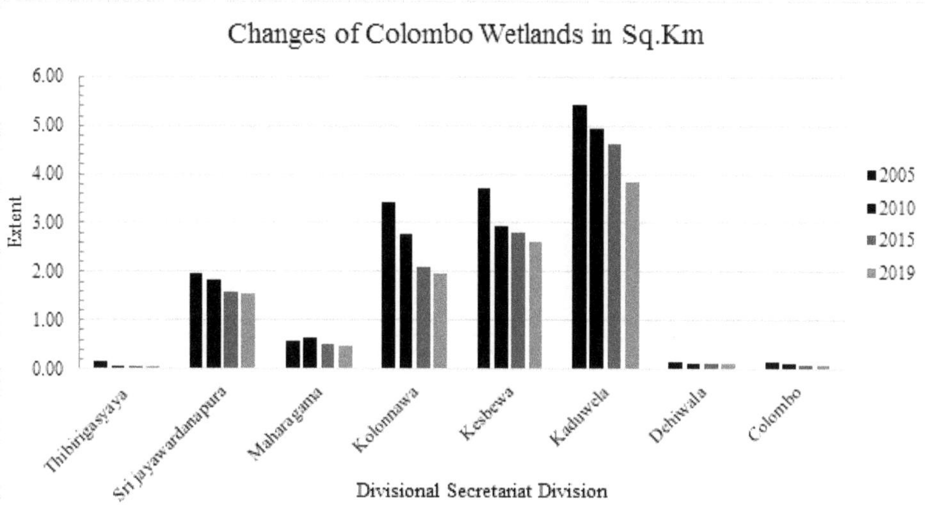

GOVERNANCE AND SUSTAINABILITY OF COLOMBO WETLANDS

From early history, wetlands were centred to originate great civilizations. Sri Lanka also has the same experience in which the Colombo wetland have been obtained greater importance in administration and power politics. In that aspect wetlands and their biota have been an integral part of the social and cultural as well as economic ethos of the people in the wetland areas since the past. This significant ecosystem has been influenced by the continuing urbanization, economic growth, industrialization and increasing population discharging more waste into wetlands (Gokce, 2018), an urban wetland in particular. Being ecologically sensitive systems they are under severe stress. Among the reasons for wetland loss and deterioration, overexploitation of the resources, land degradation, urbanization and pollution, climate change and decreasing biodiversity and invasive species are prominent (Gokce, 2018). In this sense, anthropogenic activities have both direct and indirect effects on wetlands. Notwithstanding the anthropogenic effects of the observed wetland loss and degradation in recent decades, climate change intensified impacts are more specific in wetland when it comes to the context of wetland management. Temporal variation is the basis of the degradation degree of a wetland while the ecosystem recovery and duration are dependent on two factors: the usage grade implies the decrease or increase of anthropogenic pressure; spatial and temporal variation change the carrying capacity of wetland (Gokce, 2018). Hence, feedback

Figure 6. Spatio-temporal changes of Colombo wetlands
Source: Author, 2021

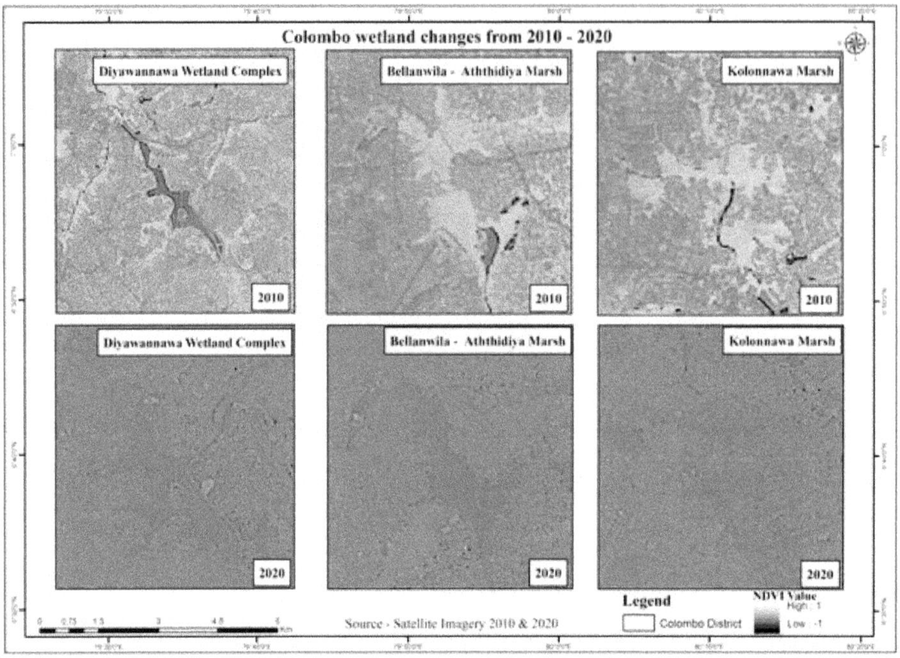

mechanisms at the wetland are crucial control systems where the wetlands are considered as holistic ecosystems in the basin-scale (Figure7). Odum (Odum, 2002) and Soto (Sorto-Ortiz, 2015) have stated that the natural equilibrium is not a steady state and has a homeorhesis. When the human population intervenes and excessive uses of wetlands, natural tolerance mechanism is destructed and wetland wanders off their homeorhesis. Approximately 87% of the wetlands in the world have been lost since 1700. The utilization of wetlands in various ways on different scales have destroyed the natural feedback process of wetlands. Therefore managing and restoring wetland ecosystems is much as important with the current trend in climate change.

Eighteen cities around the world were turned into wetlands at the 13th Wetlands Conference in Dubai, highlighting the high sensitivity of the wetlands in the Colombo District to the fact that Colombo is the only city in South Asia to be included (Mombauer, 2019). However, the frequency of heavy rainfall events has been increased in turn heightened urban flood risk which caused greater economic loss within these wetland areas.

Among the most vulnerable area for frequent flooding lower Kelani flood plain occupies an area of 1200km^2, lies through Colombo which is the commercial capital

Figure 7. Comparison a) natural wetland b) human effects on wetlands
Source: Gokce, 2018.

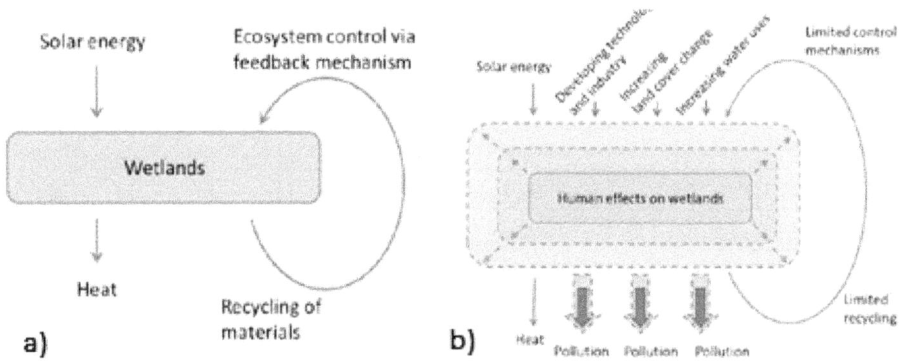

of the country belongs to the wet zone of Sri Lanka (Silva, Weerakoon, & Herath, 2016). Typical rainfall pattern has changed in the recent past and the frequency of extreme rainfall events and flash flood occurrence has increased (Table2). Hence, the flood resilience in Colombo is mainly dependent on these urban wetlands specially marshlands. It is estimated nearly 40% of Colombo floodwaters drain into these marshy areas (GoSL., 2018). Although Colombo wetlands role in flood control is significant, those spongy wetlands are decreasing rapidly at a rate of 1.2% per year. It has been estimated nearly one-third of Colombo wetlands disappear by 2038 and a half by 2070 (GoSL, 2016). If this loss continues along with the frequent urban floods and heavy rainfalls, the floodwater levels would increase by two meters (GoSL, 2016). The findings of Hettiarachchi (2018) and Missake et al (2014) have clearly emphasized the importance of marshlands and wetlands in controlling floods. Thus, this implies the necessity of preserving the wetlands and letting the natural infiltration process sustain as one major step of mitigating floods. Wetland restoration and preservation should be a must component in all development projects which are undertaken by both the public and private sectors. The wetland decline has a clear relationship with the urbanization process. In this context, urban wetland governance and related challenges to sustainable governance of Colombo urban wetlands are essential to discuss.

Since early history; throughout the colonial, post-colonial, and neo-liberal eras, the wetlands have been utilized and sometimes used for capitalist economic projects and removal of the communities dependent on them; the cycle of appropriation and dispossession. Typically this is how lands are acquired which is recently identified with the terms 'land grabbing' and 'green grabbing' (Hettiarachchi et al., 2017). Being geo-strategically important these wetlands to the British colonials they have seen wetlands as possible space for the city's expansion except for flood infrastructure.

Table 2. Flood occurrence in Kelani river basin

Year/Month	Water Level (ft. AMSL)	Category	Year/Month	Water Level (ft. AMSL)	Category
1837	13.5	Critical	1947 October	6.0	Minor
1872	11.9	Dangerous	1952 May	8.25	Major
1891	9.8	Major	1952 October	6.0	Minor
1904	9.9	Major	1955 October	8.0	Major
1906	10.8	Dangerous	1957 December	6.25	Minor
1913	11.0	Dangerous	1963 October	6.42	Minor
1922	12.6	Critical	1966 September	6.67	Minor
1925 July	11.5	Dangerous	1966 October	9.0	Major
1928 May	9.08	Major	1967 October	9.17	Major
1930 October	10.91	Major	1971 September	7.33	Minor
1933 May	9.95	Major	1975 May	6.58	Minor
1936 May	9.43	Major	1989 June	9.2	Major
1937 May	10.3	Dangerous	1999 April	6.6	Minor
1939 May	9.35	Major	2006 November	5.65	Minor
1940 May	11.0	Dangerous	2008 April	5.75	Minor
1942 July	8.17	Major	2008 May	5.9	Minor
1943 May	6.58	Minor	2011 May	5.41	Minor
1944 May	6.0	Minor	2016 May	7.64	Minor
1947 August	12.85	Critical			

Source: CRIP, 2018; Dammalage & Jayasinghe, 2019

Hence, colonials acquired and reserved large wetland parcels for flood protection which process was legally protected by the Crown land encroachment ordinance in 1840 and Wasteland ordinance in 1897. This process was continued into the period after Sri Lanka gained independence in 1948 (Hettiarachchi et al., 2017). Later the post-independence governments of Sri Lanka introduced some land-reforms policies too. Continuous population increasing increased the demand for land in the area. In line with the rising demands, urban dwellings and public infrastructure also raised. Wetland modification for the flood control programme and reclamation for new housing and road construction further altered the wetland ecology and cut off the hydraulic connectivity (Hettriarachchi et al., 2014). Neo-liberal policies introduced after 1978 started to transform Colombo into a modern commercial hub. During this period war-affected ethnic minorities migrated mainly to these Colombo wetlands or the fringes (Hettiarachchi et al., 2017). With the support of international funding, several blueprints were designed to moderate Colombo, paying special attention to slum resettlement and better utilization of the marshlands for city expansion (UNDP, 1978). This was called for a 'judicious balance between the demand for land reclamation and drainage'(Hettiarachchi et al., 2017). The initial proposal for reclaiming the wetlands for urban expansion presented in 1937 gave rise to a comprehensive proposal to reclaim the wetland for urban housing in 1966. From 1978 to 1985 >700ha of wetlands were acquired and converted for the new parliament and new capital city. Among the pioneer modernization projects, developing a new administrative capital 'Sri Jayawardanapura Kotte' was directly aimed to grab wetlands for its expansion. Consequently, clear policy drove towards reclaiming wetlands specially marshlands for urban expansion projects than ever. In 1992 second Colombo flood protection plan and wetland modification were launched under Japanese funding. Meantime several acts and law enforcing authorities such as Urban Development Authority Act, Law land Reclamation Board Act, National Environmental Act and Central Environmental Authority introduced impacting the urban wetlands of Colombo. From 1998 onwards wetlands reclamation continued for commercial projects. Notwithstanding the wetland conversion and transformation, the country made arrangements to carry out national wetland conservation studies under Netherland funding. Further hydrological modification continued under the Colombo Metro Development Plan funded by the World Bank in 2010. Along with these reforms, the Colombo wetlands were modified and the process is being on its way. The devastating loss of Colombo wetlands is clear when observing and comparing the satellite images from 1984 to 2021.

Along with this wetland transformation the land value as well as the locational rent of the areas adjoining wetlands increased. This was recently reported by Missaka et al. in 2017 where the land values increased by 110% between 1990 and 2005. In response to the increased land values, real estate in Colombo was marketed in the early 2000s and as a result, those skyscrapers and multistoried buildings became a significant feature in the modified wetland environment. In this real estate marketing process, investment and pricing had a specific relevance on wetlands. As Missaka et al., (2017) stated this relevancy or bearing on the wetlands is various; a) privately owned wetlands pressurized to be converted into marketable land; b) slums and shanties and other low-income settlements around Colombo wetlands targeted to make way for prime real estate; c) acquisition of private wetlands and altered for improving necessary drainage and road facilities; d) small residue patches of wetlands converted into waterfronts as recreational and aesthetic purposes as a means to boost real estate value in their vicinity. Some of the acquired wetlands in the new administrative city consisted of luxury apartments including golf courses, commercial complexes, and other accommodation facilities i.e. star class restaurants. The projects (03 projects) led under the financial support of Japan and World Bank mainly focused on hydrological alteration which also evacuates slums and shanties in wetlands with unauthorized construction which was seen as the main reason for flooding (Nippon Koei, 2002). In this way, the wetlands in the neo-liberal era transformed into a commodity of the urban real estate market. This massive transformation process was governed by various organizations and institutions having different wetland-related responsibilities (Figure 8).

In 2006 SLLRDC act was amended leading to acquiring nearly 500ha of private wetlands and paddy lands. After the end of the thirty-year civil war, both UDA and SLLRDC were brought under the Ministry of Defence to enforce and ensure the unhindered progress of their projects. Direct involvement of armed forces made it easy to dissipate any confrontation against the land acquisition or wetland related projects (Gunasekara, 2010). During this period urban development proposals were carefully created with meaningful words specifically environmentally sensitive language which among 'preserving wetlands as urban biodiversity refuges for recreation and ecotourism' is precise. Gradually urban wetlands became known as wetland parks; Diyatha Uyana wetland park, Beddagana wetland park, Nawala urban wetland park, Diyasaru park Thalawathugoda. Now the urban environment is really attractive but sustained hydrological modification and nutrient accumulation due to the watershed urbanization transformed marshy wetlands into shrub wetlands with a permanent reduction of the water holding capacity (Hettiarachchi et al., 2014). This is evident when comparing the two major floods that have ever happened in the flood history of the Colombo district. The recent major flood in May 2016 is the severest after the major flood in June 1989. Despite the similar magnitude of rainfall with spatial variations, the duration of inundation and the irregular development along with wetland transformation were primarily made the 2016 flood more devastating than the 1989 flood. Sri Lanka experienced the highest economic loss in South Asia due to this flood (Dammalage & Jayasinghe, 2019). Besides, the governments responded to each flood event with a new engineering project. Such projects were primarily acquired lands and evicted low-income inhabitants around the wetlands. The same areas were reconstructed with multi-story housing complexes which can be figured out using photographs (Figure9). Also, Colombo wetlands were reclaimed for transportation infrastructure; the expressway Kaduwela to Matara and the Outer circular way run over large patches of paddy and marshy lands (Gunathilaka, 2020). The upcoming construction of the highway from the new Kelani Bridge to Athurugiriya also planned to run over the Thalangama wetlands which is one of the neighbouring wetlands of Diyawanna wetland.

The unforeseen consequence of this wetland transformation is emerging hybrid wetland ecosystems. Missaka et al., (2014) emphasise that alienation or alteration of any wetland have in-depth effects on its hydrology, biodiversity, soil and totally biogeochemistry of the wetland which the long term outcome is wetland is transformed

Figure 8. Organizational structure governing Colombo wetlands
(IRR – Irrigation Department; AD – Agriculture Department; CDLLRB – Colombo District Lowland Reclamation Board; SLLRDC – Sri Lanka Land Reclamation and Development Corporation; UDA – Urban Development Authority; CEA – Central Environmental Authority; CMC – Colombo Municipal Council; DMC – Disaster Management Centre; ASB – Agrarian Services Board).
Source: Missaka et al., 2017

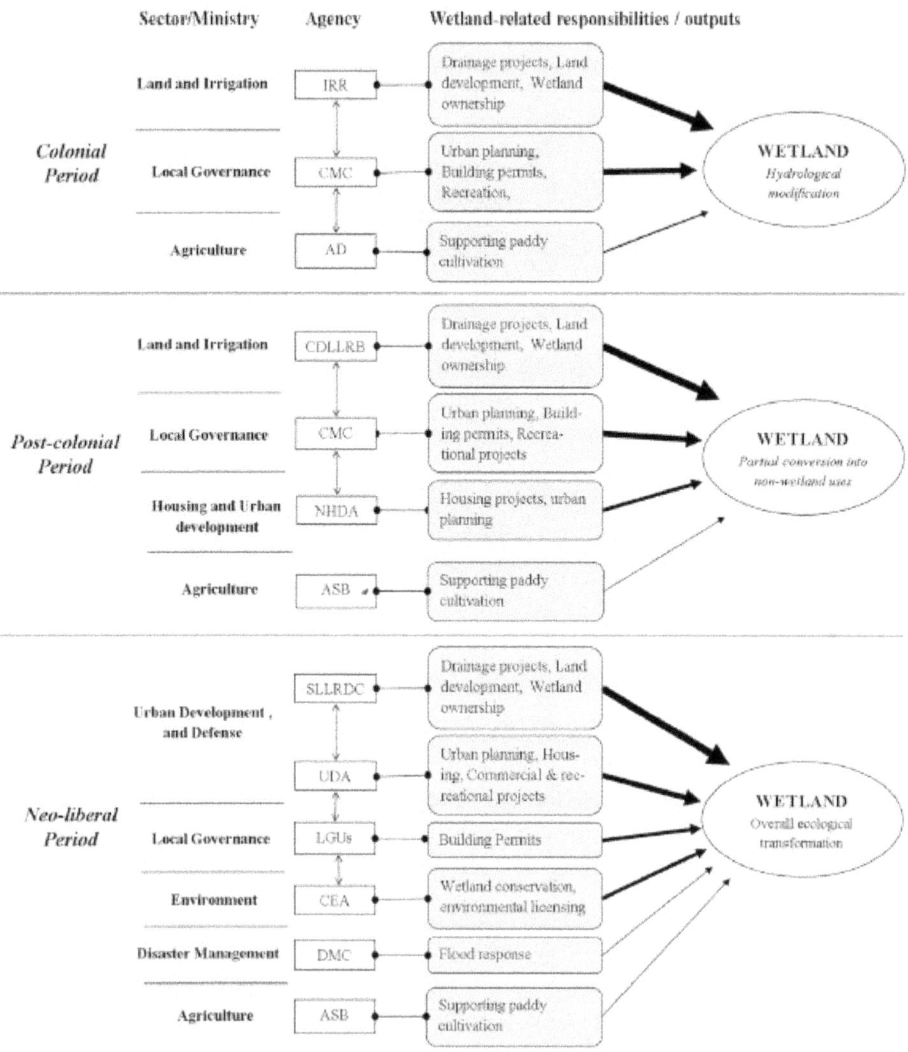

into a hybrid ecosystem. Even though this chapter does not pay attention to the socio-economic effects of wetland transformation, the transformation that takes place is truly an eco-social transformation. Since it is evident of the intensified impacts of climate change in the country (Herath & Rathnayake, 2005; ADB. 2020; Karunathilaka

Figure 9. Wetland transformation a) slums and low-income housing and emerging multistory buildings at Kotuwegoda marshes, b) Kotte marshy areas with emerged luxury apartments, c) emerging commercial hubs around Kotte Diyawannawa marshy areas
Source: Google Streetmap, 2020

et al. 2017) the importance of wetlands in Colombo is particular. Wetlands are the kidneys that filter and absorbs excess water during flooding. Besides, the loss of habitat for many species and gradual degradation of the wetlands ultimately fail the expectations of real estate customers. Therefore wetland preservation along with green growth must be the ultimate purpose of whatever development takes place within the Colombo wetland city to avoid unnecessary economic and social damage due to frequent floods.

Relying on the current scenario of urban wetlands in Colombo it is sure that the sustainability of wetlands may not be passed to the next generation. Instead deteriorated urban wetland ecosystem will remain. In this context sustainable wetland governance is essential. Urban wetlands, green growth and Sustainable Development Goals (SDGs) and the New Urban Agenda are important stepping stones of sustainable wetland governance that can contribute to several SDGs. SDG 3, 6, 11, 13, 14 and 15 are among them. Good health and wellbeing (SDG3) can be obtained by providing green spaces for leisure, education and tourism, enabling non-motorized mobility and reducing air pollution. SDG 6: clean water and sanitation through urban residents exposure to contaminated water and associated public health concerns, as well as water security through its role in flood mitigation. Sustainable cities and communities are the SDG 11 which can be achieved by providing green, liveable and environmentally sustainable urban spaces. SDG 13, 14 and 15 stand for climate action and life below water and life on land. SDG13 archives by increasing urban flood resilience and securing urban carbon pools. Eventually, protecting ecological habitats and biodiversity in urban ecosystems results in SDG 14 and 15. Hence, wetland conservation and restoration are essential under climate change risks to achieve the SDGs.

CONCLUSION

Colombo freshwater wetlands have been undergone several transformations. The spatio-temporal changes of urban wetlands demonstrated that a large extent has reduced along with the wetland governance and transformation. The intensified impacts of climate change have increased the frequency and intensity of urban floods in Colombo. Along with the Colombo wetland shrinkage the water holding capacity of wetlands has reduced and this, in turn, affects urban flood intensity. This triggered freshwater wetland shrinkage and deterioration aroused that it is time towards the freshwater ethics. Prof. Steven J. Cooke et al., (2021) reflected on Aldo Leopold's grant to recognise a set of 'Aldo-inspired' recommendations for shielding and restoring freshwater environments in an Anthropocene era through implementing an ecosystem approach; dealing with freshwaters as attached social-ecological schemes; admitting the limits to human governance over freshwater systems, and addressing the underlying causes of environmental issues rather than their warning signs. Moreover, the authors look to Leopold to recap us that freshwaters are a central component of varied human-nature affairs, and a sense of 'freshwater optimism' is compulsory to make optimistic conservation progress (Steven et al., 2021). Freshwater ecosystems are critical to our survival – and Leopold reminds us not only of our ethical responsibility towards them but that our fate will be tied to theirs (Freshwaterblog, 2021). Colombo being a wetland city, green grabbing should be more focused to achieve SDGs and ensuring the sustainability and surveillance of wetlands to tackle the climatologically intensified urban floods. Otherwise, socio-ecological transformation altered the wetland ecosystem ad ultimately resulted in a hybrid ecosystem. To prevent this unforeseen tragedy this study suggests introducing freshwater ecosystem ethics to Colombo wetlands to ensure the sustainable survival of the ecosystem and application of advanced restoration and conservation methods to protect Colombo wetlands

REFERENCES

ADB. (2020). *Climate risk country profile: Sri Lanka*. Asian Development Bank.

Bambaradeniya, C. (2003). Freshwater wetlands in Sri Lanka: their conservation significance and current status. *National symposium on wetland conservation and management* (pp. 19-24). Colombo7: IUCN.

CEA. (1993). *Wetland site report and conservation management plan*. CEA.

Constanza, R., d'Arge, R., DeGroot, R., Farbe, S., Grasso, M., Hannon, B., & O'Neill, R. V. (1997). The value of the worlds ecosystem services and natural capital. *Nature*, *387*(6630), 253–260. doi:10.1038/387253a0

Crandell, J. (2021). Geographic Distribution of the wetlands. *Britannica*. https://www.britannica.com/ science/wetland/Geographic-distribution-of-wetlands

CRIP. (2018). *Strategic Environmental Assessment of Development of River Basin Level Flood and Drought Mitigation Plans*. Kotte: Consulting Engineers & Architects Associated (Pvt.) Ltd. http://crip.lk/wp-content/uploads/2019/01/Kelani-River-SEA-Study-Final-Report.pdf

Dammalage, T., & Jayasinghe, N. (2019). Land-Use Change and Its Impact on Urban Flooding: A Case Study on Colombo District Flood on May 2016. *Engineering, Technology &. Applied Scientific Research*, *9*(2), 3887–3891.

Dayathilaka, D., Lokupitiya, E., & Wijerathne, V. (2020). Estimation of aboveground and belowground carbon stocks in urban freshwater wetlands of Sri Lanka. *Carbon Balance and Management*, *15*(17), 17. doi:10.118613021-020-00152-5 PMID:32876789

DWC. (2016). Vanadivi theth bim. *Science.gov*.

Fernando, P. (2010). Disaster management measures of microfinance. *Sri Lanka Journal of Advanced Social Studies*, *2*, 30.

Finlayson, C., Milton, G., & Prentice, R. (2016). Wetland types and distribution. In C. Finlayson, G. Milton, R. Prentice, & N. Davidson (Eds.), *The wetland book*. Springer. From., doi:10.1007/978-94-007-6173-5_186-1

Freshwaterblog. (2021). The freshwater blog is the voice of aquatic life. *Fresh Water Blog*. https://freshwaterblog.net/2021/03/10/towards-a-freshwater-ethic-lessons-from-aldo-Leopold-for-contemporary-aquatic-conservation/

Gokce, D. (2018). Wetland importance and management. In *Wetland management-Assessing risk and sustainable solutions* (pp. 1–9). Intech Open.

GoSL. (2016). *Final Metro Colombo Wetland Management Strategy.* Landscape and urban design office.

GoSL. (2018). *State of Sri Lankan Cities*. UN-Habitat.

Gunasekara, T. (2010 May 12). The development war-a 'humanitarian operation' against Colombo's poor. *The Sunday Leader*.

Gunathilaka, M. D. K. L. (2020). Land use and land cover changes and Avifauna: An empirical analysis of loss of agricultural wetlands and its impact on avian species in suburban areas. *International Journal of Scientific and Research Publications, 10*(5), 263–272. doi:10.29322/IJSRP.10.05.2020.p10132

Henkanaththegedara, S. (2005). Ichthyofauna of its Bellanwila-Attidiya sanctuary and its environs in Colombo, Sri Lanka. *TriggerPaper, 32*(1).

Herath, S., & Rathnayake, U. (2005). Changing rainfall and its impact on landslides in Sri Lanka. *Journal of Mountain Science*, *2*(3), 2018–2224.

Hettiarachchi, M., Athukoral, K., Wijekoon, S., & Alwis, A. (2014). Urban wetlands and disaster resilience of Colombo, Sri Lanka. *International Journal of Disaster Resilience in the Built Environment*, *5*(1), 79–89. doi:10.1108/IJDRBE-11-2011-0042

Hettiarachchi, M., Morrison, T., & McAlpine, C. (2017). Power, politics and policy in the appropriation of urban wetlands: The critical case of Sri Lanka. *The Journal of Peasant Studies*, 1–19.

Hettiarachchi, M., Morrison, T., Wickramasinghe, R., Mapa, A., & McApine, C. (2014). The eco-social transformation of urban wetlands: A case study of Colombo, Sri Lanka. *Landscape and Urban Planning*, *132*, 55–68. doi:10.1016/j.landurbplan.2014.08.006

Hettiarachchi, P. (2018). Flood frequency analysis and selection of suitable design flood for Kelani flood bunds. *Research Gate.* https://www.researchgate.net/publication/327745195_Flood_Frequency_Analysis_for_the_Kelani_River_and_the_Design_Flood_for_Kelani_Flood_Bunds

Hewavitharana, B. (2004). Poverty alleviation. In S. Kelegama (Ed.), *Economic policy in Sri Lanka: Issues and debates* (pp. 467–497). Sage Publication.

IUCN. (2006). *CEA*. National Wetland Directory of Sri Lanka.

Kabii, T. (n.d.).An overview of African wetlands. *Aquadocs.* https://aquadocs.org/bitstream/handle/1834/457/Africa_Wetlands Retrieved Dec. 9. 2021.

Karunathilaka, K., Dabare, H., & Nandalal, K. (2017). Changes in rainfall in Sri Lanka during 1966-2015. *Journal of the Institution of Engineers, Sri Lanka, 50*(2), 39–48. doi:10.4038/engineer.v50i2.7251

Kawamura, M., Kominami, K., & Takeuchi, J., & A., T.-e. (1996). A multicopy suppressor of nin1-1 of the yeast *Saccharomyces cerevisiae* is a counterpart of the Drosophila melanogaster diphenol oxidase A2 gene, Dox-A2. *Molecular & General Genetics, 251*(0), 146–152. PMID:8668124

Koei, N. (2002). *The study on stormwater drainage plan for the Colombo metropolitan region in the democratic socialist republic of Sri Lanka.* Sri Lanka Land Reclamation and Development Corporation, JAICA.

Kotagama, S. W., & Bambaradeniya, C. B. (2006). An Overview of the Wetlands of Sri Lanka and their Conservation Significance. In National Wetland Directory of Sri Lanka (pp. 7-16). The Central Environmental Authority (CEA), The World Conservation Union (IUCN),the International Water Management Institute (IWMI).

Marawila, T., & Tibbotuwawa, M. (2010). *To develop or to conserve: the case of the Diyawanna Oya wetlands in Sri Lanka.* South Asian Network for Development and Environmental Economics.

Matthews, G. (2013). The Ramsar Convention on wetlands: its history and development (re- issue of 1993 edition ed.). Ramsar Convention Bureau.

MCUDP. (2016). *Metro Colombo Urban Development Project.* SLLRDC.

MoEF. (2020). A conceptual model to understand the drivers of change in tropical wetlands: A comparative assessment in India and Brazil. *Biota neotropical.* doi:10.1590/1676-BN-2019-09-13

Mombauer, D. (2019). Colombo's urban wetlands: on becoming and staying a Ramsar wetland city. *Slycan Trust.*

Odum, E. (2002). Tidal marshes as out welling/pulsing systems. In M. Weinstein & D. Kreeger (Eds.), *Concepts and Controversies in Tidal Marsh Ecology* (pp. 3–7). Kluwer Academic. doi:10.1007/0-306-47534-0_1

Panditaratna, B. (1965). The trend of industrialization in Colombo city: the Capital of Ceylon. *Pakistan Geographical Review, 20*(2).

Pathirana, A., Hailu, B., William, V., Chris, Z., & Allan, B. (2014). Impact of urban growth- driven land-use change on microclimate and extreme precipitation - A sensitivity study. *Atmospheric Research*, *20*(2), 59–72. doi:10.1016/j.atmosres.2013.10.005

Postel, S. (2019). Recent Loss of Freshwater Wetlands Worldwide Valued at 27 Trillion a Year. *National Geographic*. https://blog.nationalgeographic.org/2014/06/24/recent-loss-o f-freshwater- wetlands-worldwide-valued-at-2-7-trillion-per-year

Ranasinghe, P. (2018). *Ramsar accreditation of Colombo Wetlands: Rebranding Colombo as a wetland city*. Daily FT.

Ranasinghe, P., Piyadasa, R., & Mataraarachchi, S. (2019). A study to identify the soil characteristics of wetlands in Colombo, Sri Lanka. Dubai: *13th Wetland Conference*.

Reddy, K. R., & DeLaune, R. D. (2008). *Biogeochemistry of Wetland*. Tyler and Frances group.

Samarasinhe, Y., & Dayawansa, N. (2013). Remote sensing and GIS-based study in an assessment of the degradation risk of the Kolonnawa marsh. *Journal of the National Science Foundation of Sri Lanka*, *41*(4), 327–335. doi:10.4038/jnsfsr.v41i4.6256

Scott, D. (1989). *A Directory of Asian Wetlands*. IUCN.

Shine, C., & Klemm, De. C. (1999). Fisheries conservation and management and the conservation of marine biological diversity in developments in international fisheries law. Kluwer.

Silva, G., Weerakoon, S., & Herath, S. (2016). Event-Based Flood Inundation Mapping Under the Impact of ClimateChange: A Case Study in Lower Kelani River Basin, Sri Lanka. *Hydrology: Current Research*, *7*(1). doi:10.4172/2157-7587.1000228

Sorto-Ortiz, L. (2015). The regulation of ecological communities through feedback loops: A. *Research in Zoology*, *5*, 1–15.

Sri Lanka Land Reclamation and Development Corporation. (2015). *Technical report 2 Ecological status*. [Data set]. Sri Lanka Land Reclamation and Development Corporation.

Steven, J., Abigail, J., John, J., Julian, D., Andrea, J., & Steve, J. (2021). Stewardship and management of freshwater ecosystems: From Leopold's land ethic to a freshwater ethic. *Aquatic Conservation*, 1–13.

Subasinghe, S., Estoque, R., & Murayama, Y. (2016). Spatiotemporal analysis of urban growth using GIS and Remote sensing: A case study of the Colombo metropolitan area, Sri Lanka. *International Journal of Geo-Information*, *5*(197), 197. doi:10.3390/ijgi5110197

Sulthana, P., & Thompson, P. (2010). *Natural resource conflicts and community organizations in Bangladesh*. Department of Fisheries.

Sushmita, S. (2012). Concrete invasion: Colombo's marshes cannot avert floods the way they used to. Construction is to be blamed. *Down to Earth*. https://www.downtoearth.org.in/coverage/urbanisation/concrete-invasion-35850

UNDP. (1978). *Colombo master plan project*. Ministry of Local Government & Housing & Construction, Department of Town & Country Planning.

UNFCC. (2018, October 1). Wetlands disappear three times faster than forests. *United Nations Framework Convention on Climate Change*. https://unfccc.int/news/wetlands-disappearing-three-times-faster-than-forests

UoM. (2011). Environmental screening report for establishment of Beddagana and Kotte Ramparts Parks. *Sri Lanka - Metro Colombo Urban Development Project: environmental assessment*, 8.

Wetzel, R. (2001). *Limnology: Lake and river ecosystems* (3rd ed.). Academic Press.

Zon, V. (2003). Wetlands conservation and management in Sri Lanka: A status paper. *National symposium on wetland conservation and management* (pp. 2-18). IUCN.

ADDITIONAL READING

Acreman, M., & Holden, J. (2013). How Wetlands Affect Floods. *Wetlands*, *33*(5), 773–786. doi:10.100713157-013-0473-2

Erwin, K. (2009). Wetlands and global climate change: The role of wetland restoration in a changing world. *Wetlands Ecology and Management*, *17*(1), 71–84. doi:10.100711273-008-9119-1

Groot, R., Stuip, M., Finlayson, M., & Davidson, N. (2006, November). *Valuing wetlands: Guidance for valuing the benefits derived from wetland ecosystem services*. Ramsar Convention Secretariat.

Horowitz, Leah. (2013). Toward Empathic Agonism: Conflicting Vulnerabilities in Urban Wetland Governance. *Environment and Planning A: Economy and space, 45*, 10.

MEA. (2005). *Ecosystems and human well-being: wetlands and water.* World resource institute.

Mishra, N. (2020). Wetlands: Remote sensing. In Y. Wand (Ed.), *Wetlands and habitats*. Routledge., doi:10.1201/9780429445507-27

Simpson, G., & Newsome, D. (2017). Environmental history of an urban wetland: from degraded colonial resource to the nature conservation area. *Geography and Environment, 4*(1).

Verhoeven, J., Beltman, B., Whigham, D., & Bobbink, R. (2006). Wetland functioning in a changing world: implications for natural resource management. In J. Verhoeven, Beltman., Bobbink., & Whigham (Eds.), Ecological studies (Vol. 190). Springer. doi:10.1007/978-3-540-33187-2_1

KEY TERMS AND DEFINITIONS

Anthropogenic Climate Change: Humans area unit inflicting the foremost of current changes to climate as well as heating and emissions of greenhouse gases.

Green Grabbing: An oversized scale personal appropriation of land, resources, and water legitimized with the protection of the setting or supported through mechanisms associated with global climate change mitigation. Almost like the thought of land grabbing it's a political term, used each by activists and scholars to criticize massive scale land appropriation processes. green grabbing or land grabbing is the privatization or appropriation of land for functions of advancing a "green" economy whereas excluding native, endemic individuals from natural resources.

Freshwater Ethics: Elaborating our values and moral principles regarding freshwater and nature, and concerning individuals and cultures, facilitate to grasp higher who we are, however, individuals think of the planet and the way their neighbours do an equivalent. A field of freshwater ethics, with space for divergent however ethically grounded views, will facilitate people to understand themselves and others deeply and realize new and surprising solutions to the challenges of the freshwater ecosystems.

Integrated Freshwater Landscape: Spatially connected lakes, wetlands, and rivers to at least one another at their macro-scale.

Social-ecological Transformation: An umbrella term that describes recent political, socioeconomic, and cultural shifts ensuing from attempts to handle the social-ecological crisis.

Wetland Ecosystem Services: Direct and indirect products and services provided by the wetlands employing a dynamic and broad community.

Wetland Governance: A method characterised by multi-stakeholder administration with dynamic diplomacy among native governments over wetlands to rework them into commercially necessary financial assets.

Wise-use of Wetlands: Maintenance of wetland ecological characters, achieved via the appliance of ecosystem approach among the context of sustainable development.

Section 3

Groundwater and Marine Biodiversity in a Context of Climate Change

Chapter 4
Impact of Climate Change on Subterranean Wetland Biodiversity in Tafilalet South-Eastern Morocco

Asma El Alami El Filali
Faculty of Sciences Smelalia, University Cadi Ayyad, Morocco

Research Center BIOBIO, Faculty of Sciences, Mohammed V University, Rabat, Morocco

Abdelkrim Ben Salem
https://orcid.org/0000-0002-2283-5928

Mohammed Messouli
Faculty of Sciences Semlalia, Univeristy Cadi Ayyad, Morocco

ABSTRACT

Groundwater in Tafilalet supports diverse faunas (stygofauna) that include many obligate groundwater-dependent species (stygobites). Prospecting campaigns on underground fauna in Tafilalt region were collected in a database that was later used to develop distribution patterns for each species harvested. Tafilalet is characterized by low rainfall, high temperatures, and very high evaporation. Those severe climate conditions influence water availability for vegetation growth and fauna stability. The present work aims to, assess stygobiont richness in subterranean environment of Tafilalet. Species data were organized in database form and geographic information system (GIS) was used to establish geographical patterns of species and test fauna and environment parameters relationship, to identify a network of stygofauna sites priority for conservation and defined areas with high taxonomic richness and high level of endemism. Risk assessment and decision-making framework for managing groundwater-dependent ecosystems (GDEs) development are required.

DOI: 10.4018/978-1-7998-9289-2.ch004

INTRODUCTION

In the south of Morocco, the groundwater of Tafilalet region supports several interesting species richness. In recent years, the region has experienced overexploitation of its natural resources; the succession of long periods of drought and rainfall not exceeding 120 mm/year, and the impact of climate change, on one side, and demographic pressure and the effect of human activities, particularly since the construction of the Hassan Eddakhil dam, which has dramatically reduced the piezometric level of the water table and its biodiversity and has also increased the degree of groundwater pollution.

From another side, they have contributed to the change in the biodiversity of underground aquifers and to the increase in the degree of endemism of species, sometimes even to their extinctions. Permanent environmental degradation affects the stability of underground aquatic habitats. As a result, many stygobiont species are threatened the higher their degree of endemism, the more threatened they are. In addition, ongoing global changes (climate change, drought, floods, deforestation, habitat destruction and the spread of alien and invasive species...) that are extremely rapid and human-induced continue to impoverish biodiversity. These problems are affecting many parts of the region called "hot spots", where the frequency of occurrence is greater, the sensitivity is higher; like springs of Meski and Zoual which research proved their specific richness (Messouli et al., 2009). The devastation is more severe, and local communities are more vulnerable. Water resources are a crucial area of vulnerability in Tafilalet, affecting water supply, agriculture, pasture, forest, and others.

Since the 1970s, several Moroccan zones have experienced a general rainfall decline (Driouech, 2009; Khomsi, 2015). For the Tafilalet Oases, groundwater and surface water are interconnected resources. This connectivity can have significant implications for both water availability and quality and presents major challenges for water managers and policymakers (Messouli et al., 2009). In this area, a typical gradient can be observed from humid/sub-humid subtropical mountains and sandy deserts. In the Oro-Mediterranean (mountainous Mediterranean) zone, between 2600 and 3400 m which coincides, with the most important zone for karstic groundwater creation, a large-scale degradation of vegetation has occurred. Regarding the climatic conditions, Tafilalet is characterized by low rainfall, high temperatures, and evaporation. Those severe climate conditions influence the availability of water for vegetation growth and fauna stability. The present work aims to, firstly, assess stygobiont richness in subterranean environment of Tafilalet. Species were organized in database form and projected by using geographic information system (GIS) to establish geographical patterns of species to test relationship between fauna and environment parameters, to identify a network of stygofauna sites priority for conservation and defined areas with high taxonomic richness and high level of endemism. Lately, to develop and test a risk assessment and decision-making framework for managing groundwater-dependent ecosystems (GDEs) with declining water levels due to climate change, anthropogenic extraction, land use and land management.

BACKGROUND

Over the last few centuries, the economic development that the world has experienced as well as the demographic explosion and the migration of populations, have disrupted the natural functioning of ecosystems. The impact of the dysfunction of ecosystems is reflected in the permanent reduction of their species richness. Currently, the richness of biodiversity is in continuous and unprecedented decline. Although species extinction has always been a natural phenomenon, the rate of extinction has increased dramatically by at least a factor of 100-1000 over the natural rate due to human activities (Vitousek et al., 1997; Pimm and Lawton, 1998; Koh et al., 2004). Among the ecosystems that have suffered these alterations are aquatic ecosystems and more specifically, groundwater (GW). GW not only constitute a vast reservoir of water, but they are also, in a much more difficult to perceive way, the reservoir of an original biological diversity composed of organisms, most of which have no equivalent in surface ecosystems and some of which are tens of millions of years old (Boutin, 1993; Messouli, 1994; and Humphreys, 2000).

Organism's communities (from protozoa to vertebrates) that colonize underground environments (karstic or porous) are very heterogeneous, composed of variable proportions of organisms from surface waters (epigeal species) and organisms from underground waters (hypogeal species). Species can be found (Gibert, Danielopol, and Stanford, 1994):

1. Stygobionts are species which morphologically, physiologically and/or ethologically adapted to groundwater and realise their entire life cycle exclusively in this environment.
2. Stygophiles correspond to the epigean species having a certain affinity with the underground aquatic domain. They exploit its resources and carry out partially or completely their vital cycle.
3. Stygoxens correspond to typically epigean species which don't have affinity with the underground aquatic domain, but they can be encountered there accidentally.

In Africa, research on subterranean fauna has been progressing in many countries especially in North Africa (Chappuis, 1953; Deboutteville, 1960; Boulanouar, 1982).

Moroccan underground aquatic fauna was in 1980 an almost unexplored field of research. Thirty years before, only half a dozen Crustacea were collected by Professors Rémy and Panouse; one in spring of the High Atlas and the others in the alluvium of the Oued of southern Morocco. In the region of Tafilalet, research of interstitial fauna in Oued Gheris valley was undertaken by Ben Azzouz (1983) as well as the study of the subterranean fauna of Tafilalt was carried out by Boughrous (2007).

OVERVIEW OF THE STUDY AREA

Tafilalet Environment

Characterized by its geographical location between the southern Atlas latitudes 29°30°N and 32°30N in south-eastern Morocco and spread over an area of 60,000 km^2, Tafilalet region offers a specific biological richness in its underground aquifers (Figures 1).

Depending on the geological context: The lithology present in the study area is of age: Paleozoic, Triassic, Jurassic, Cretaceous, Tertiary and Quaternary, from which the main groundwater and deep layers are formed.

The area is characterized by a semi-desert Saharan climate with a strong continental influence. Annual average temperature varies between 15°C to 22°C and can reach 50°C in Rissani during summer. Rainfall varies between 50 mm and 250 mm, with an average of 120 mm. Evaporation is very strong and varies between 2500 to 5000 mm/year. Prevailing winds are from the North-East direction.

Surface water comes mainly from the mountain area of the High Atlas, the volume of the contributions is estimated at 656 Mm^3. Groundwater in the region is also important, the renewable water reserves are estimated at 204 Mm^3/year (figure 2).

Data Collection and Database

The database allows the storage, management, analysis, and visualization at different spatial scales of a large amounts of data which have interesting not only to taxonomists and systematists (Alroy, 2002, 2003; Meier and Dikow, 2004; Sihvonen and Siljander, 2005) but also to ecologists and biogeographers (Dray, 2003). Among these data, species identification and geographic distribution are certainly the most important information (Stork and Samways, 1995; Soberon et al., 2000).

Prospecting campaigns results about stygofauna in Tafilalet, from the first campaign in 1981 until the last results available in 2021, was collected in a database then used to elaborate the distribution patterns of every group. Each record corresponds to a species identity card (Group, Genus, Family and Species) containing information about the location (sampling site, georeferenced coordinates: latitude and longitude, physicochemical characteristics, bibliographic origin ...).

Figure 1. Location of the Tafilalet region
Source: (Authors)

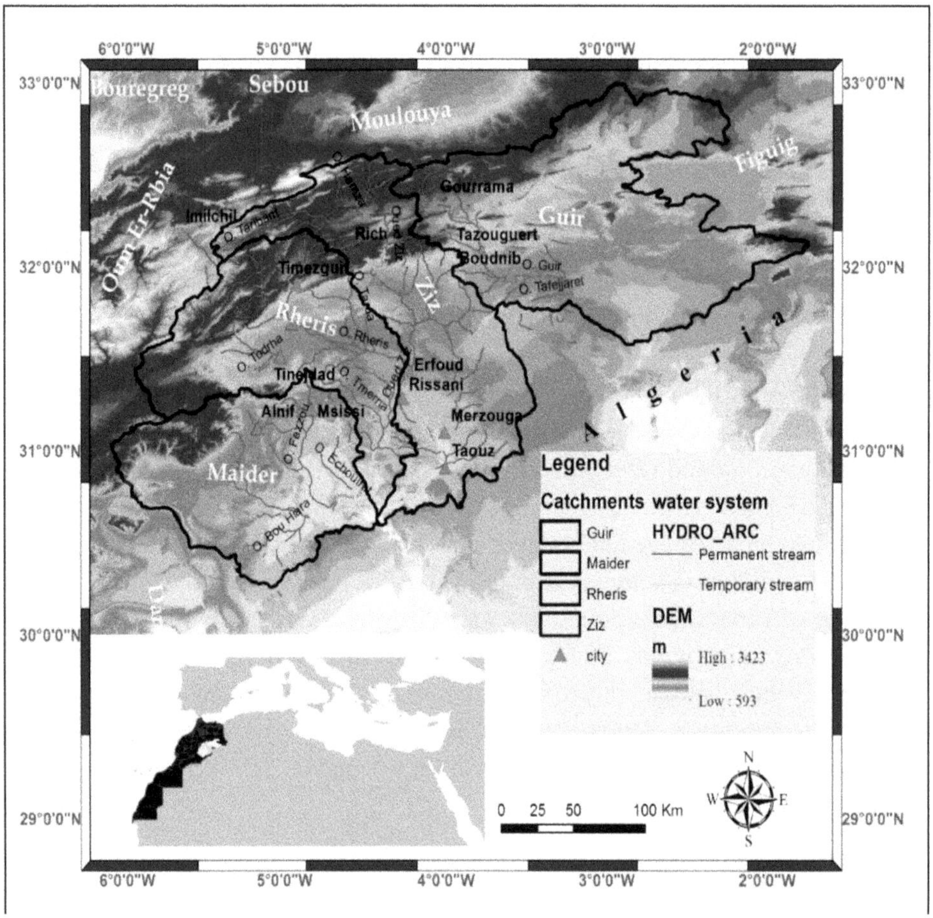

DATA ANALYSIS: STYGOFAUNA RICHNESS OF TAFILALET

To develop strategies for conservation of biodiversity groundwater in Tafilalt scale, it was necessary to identify patterns distribution of species by projecting information already stored in the database form on the map background of the watersheds of the Tafilalt region, using Geographic Information System (GIS) software. This tool is used also, to calculate the density of points. The Point density tool calculates the density of point features around each output raster cell. Conceptually, a neighborhood is defined around each raster cell center, and the number of points that fall within the neighborhood is totaled and divided by the area of the neighborhood (Silverman, 1986).

Before the first research on Moroccan stygofauna in 1980, the Tafilalet region was fortunate to be among the first sites prospected. During more than a quarter century of research on its subterranean aquatic aquifers; a specific richness represented by new families, genus and species that have been discovered. However, studies according to all these Moroccan discoveries have certainly contributed to better understand the origin of this fauna, which resulting to the" biological processes in space and time: colonization, evolution, and speciation of interstitials stygobionts widely used on worldwide scale (Boutin and Coineau, 1990). Unfortunately, stygofauna richness still largely underestimated both taxonomically and spatially; taxonomically because several species remain undetermined and spatially because some regions are not yet prospected. The database contained 27 species corresponding to more than 83 records (63%, 19% and 18% were from wells, underflows, and springs, respectively) (figure 3). These species were belonged to three groups with 69%, 19% and 12% corresponded to crustaceans, mollusks, and annelids, respectively (figure 4). Species were distributed among 14 families and 17 genus (Table 1).

PATTERNS OF GEOGRAPHICAL DISTRIBUTION OF STYGOFAUNA IN TAFILALET

According to obtained results (Figures 4, 5, and 6), we note that Amphipoda, Isopoda and Mollusk were widespread throughout the region in comparison to Oligochaeta that some new families and new species are discovered yet described which approve that research should be focused on their subterranean reserves (Figure 6a). However, Thermosbaenacea as group that colonize underground habitats and sometimes coasts but often in continents. This group has a low dispersal capacity, hence its biogeographical interest (Boutin 1993). They were only represented by one species *Monodella atlantomaroccana* belonged to Monodellidae *Family,* issued from one site (Figure 6b) which increases its endemism and its risk to disappearance especially with the impact of anthropogenic stress exerted on subterranean system like the over extraction of water for human needs, water pollution and climate change. Moreover, Isopoda species belonged to three families Microparasellidae, Cirolanidae and Microcerberidae represented eveyone by a single species (Figure 6c). It is imperative to note the discovery of two new species of *Proasellus* genus and one new species of *Thyphlocirolana;* both are undescribed. Concerning Amphipoda group which represented the largest distribution at scale area and biodiversity in

Figure 2. Localization of five stygobie species groups (Oligochaeta, Metacrangonyx, Isopodea, Amphipodea and Gasteropodea) in Tafilalet zone
Source: (Authors)

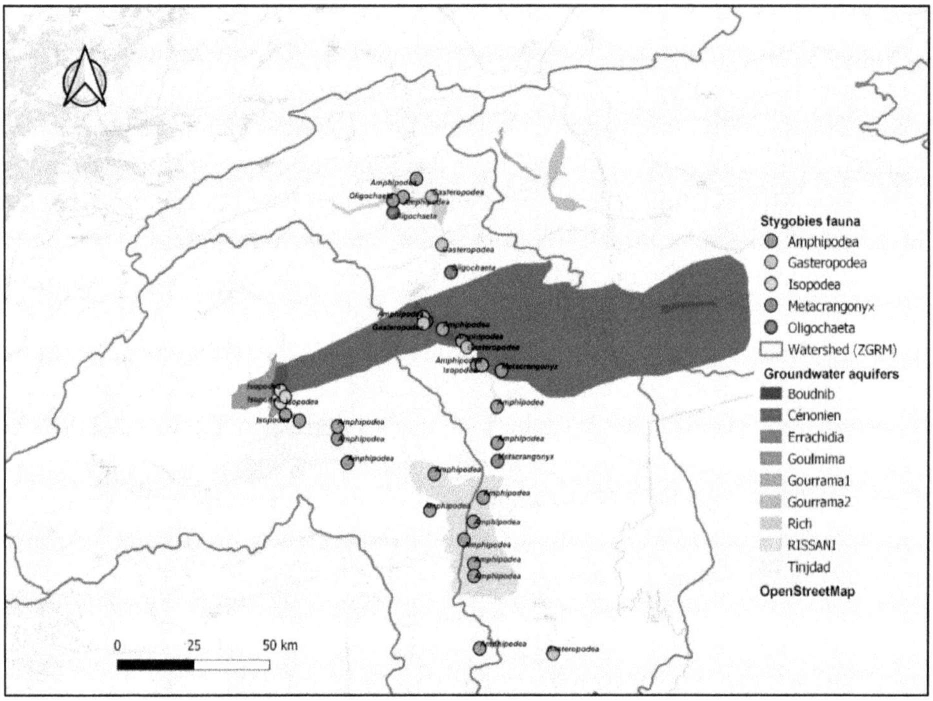

Table 1. Number of families, genus and species which was harvested from subterranean aquifers of Tafilalet

Taxa	Family	Genus	Species
Annelida			
Oligochaeta	2	3	3
Mollusc			
Gasteropoda	1	3	7
Crustacea			
Syncarida	1	1	1
Isopoda	3	3	5
Amphipoda	5	6	13
Total	12	16	29

Figure 3. Relative percentage of sampling sites of stygofauna of Tafilalet (W: well, S: spring and underflow)

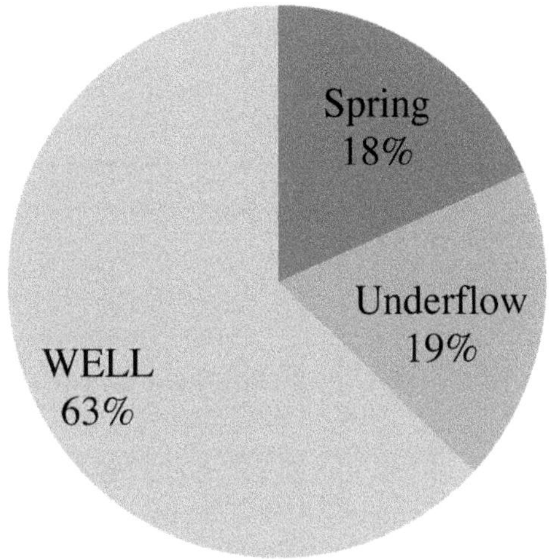

Tafilalet and have been collected from all types of habitats (underflows, wells, springs) and rarely caves. (Figure 6d). This are more studied by taxonomist, but also new species belong to *Sallentinella, Gammarus, Pseudoniphargus, Maghrebidiella* and *Metacrangonyx* were discovered but undescribed yet. Finally, the subterranean Gastropoda, currently identified, belonged to unique family of Hydrobiidae. In this region, the species collected belong to two genera: Guistia and Heideella (Figure 6-e) with the discovery of new genus which endemic to Tafilalet region.

Figure 4. Major stygobiont taxa and composition of stygobiont groups identified in Tafilalet

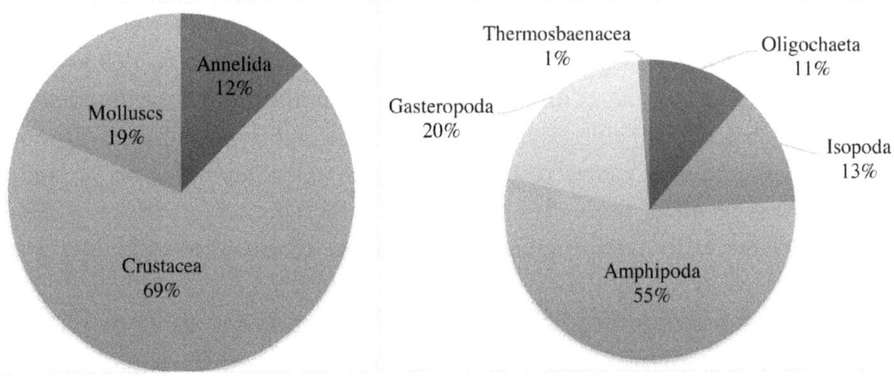

Figure 5. Composition of stands sampled in springs and wells in the Tafilalet region (genus and species classified by decreasing frequency)
Source: (Author)

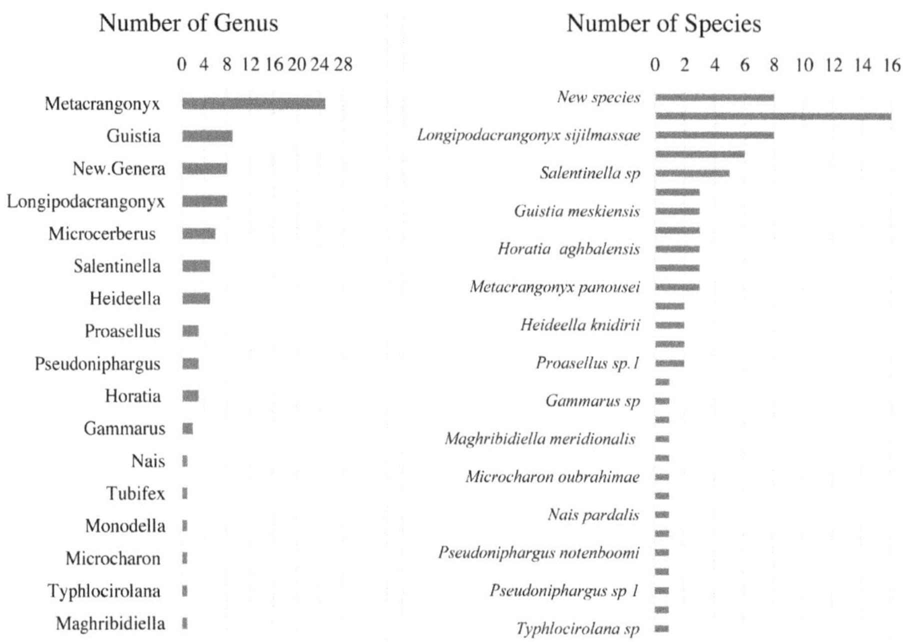

In the following, these distribution patterns of stygofauna richness attempted to characterize sites contained a high biodiversity like "Zouala" which can classified as a "hotspot" site that must be preserved from ecological point of view.

IMPACT OF CLIMATE CHANGE ON GROUNDWATER AND BIODIVERSITY

In recent years, Tafilalet has experienced a natural resources scarcity due to naturel factors such as, long periods of drought succession, annual rainfall scarcity that does not exceed 120 mm/year, climate change impact and, anthropogenic activities over their overexploitation of natural resources, increase in the number of people, dams' building. As a result of these changes a remarkable decrease in the piezometric level (PL) of the water table (Figure 7). This figure clearly shows the drop in the piezometric level of two aquifers, which host a great specific richness of the stygobia fauna. The piezometric level of the Fezna Jorf aquifer moved from -12 m in 1954 to -18 m deep in 2016, likewise the (PL) of the Lhain Tizimi aquifer moved from -7 m

Figure 6. Density and patterns of distibution of the five zoological groups (Taxa)

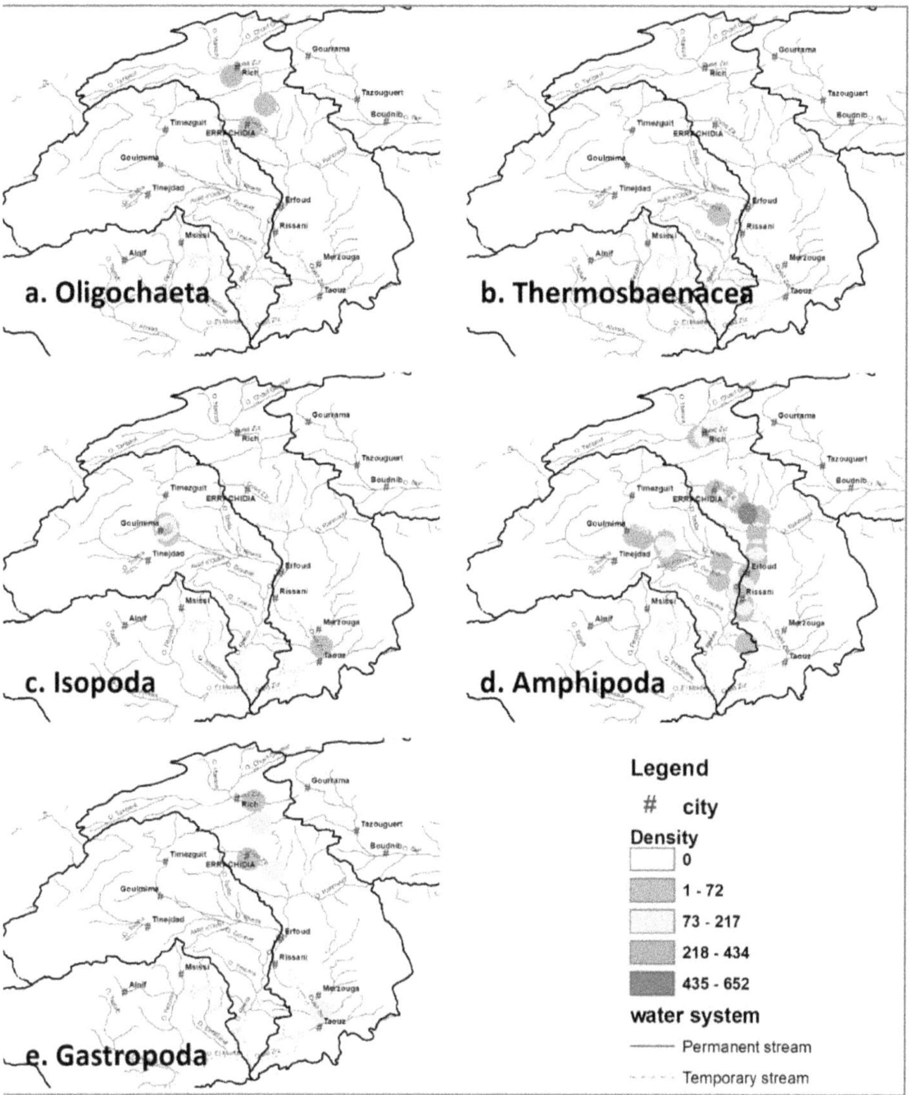

in 1954 to -9 m in 2016 with fluctuations interannual periods marking the specific climatic variability of arid zones. This change in PL is directly reflected on wells, springs, and underflows, therefore, in a direct way it will influence their groundwater quantity and quality and consequently stygobia fauna quality and quantity. Decrease of biodiversity accompanied by an increase in the degree of endemism of species and sometimes even their extinction.

Figure 7. change in the piezometric level of the Fezna Jorf (N°IRE 1028/57) and Lhain Tizimi-Erfoud (N°IRE 454/57) water table between 1954 and 2016
Source: (ABH GZR 2017)

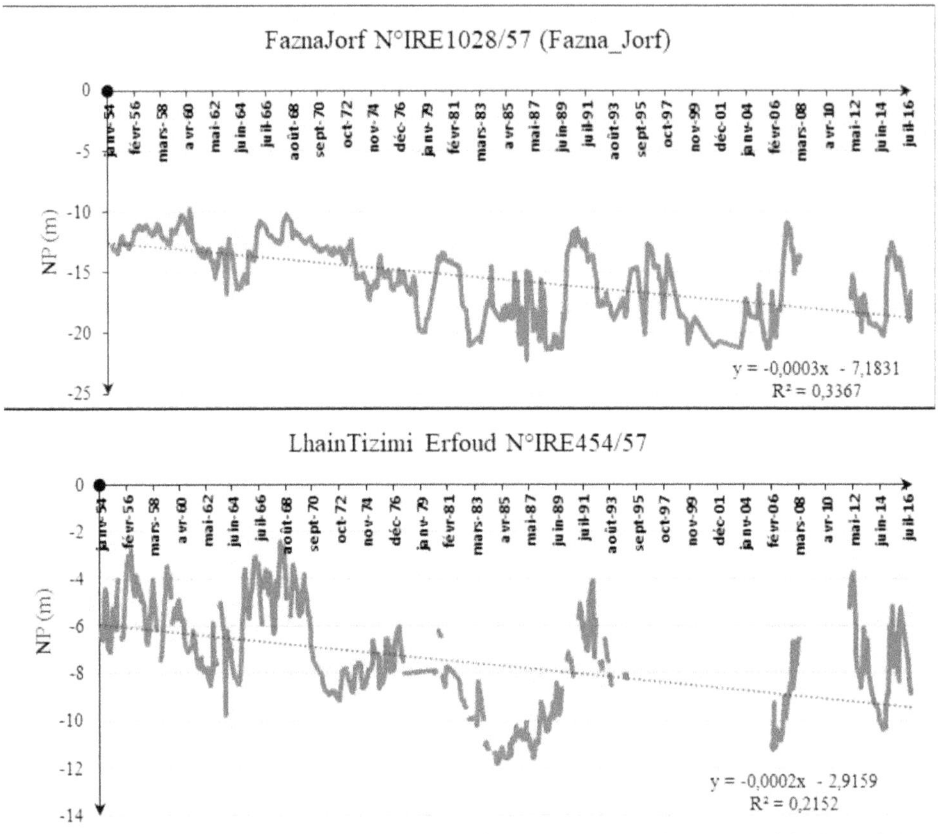

Aquatic subterranean species are often endemic species. Therefore, they are both a part of the national natural patrimony, and a significant part of the regional natural patrimony to consider when conservation measures are planned. The diversity of the aquatic fauna occurring in wells and springs of a given region is maximum when water is of good quality and decreases in case of local water pollution. Stygobiontic species are generally more sensitive to pollution than surface freshwater species, and they include indicative species of water quality. The knowledge and comparisons of the diversity of faunal assemblages from wells, springs and even from drill holes from an arid region can be used for monitoring purpose, protection, and management of the environment, and pointing to local water pollutions) they can be useful for public health (Boutin et al., 2011).

Conservation Implications

Continued environmental degradation affects the stability of underground aquatic habitats. As a result, many stygobian species are threatened the more endemic they are. In addition, ongoing, extremely rapid, human-induced global changes (climate change, drought, flooding, deforestation, habitat destruction, and the spread of exotic and invasive species) continue to deplete biological diversity. To cope with these changes, some species develop adaptation strategies as is probably the case of *Proasellus coxalis africanus*, which has just colonized groundwater following the drying up of surface rivers; this is the phenomenon of "stygogenesis" (Boulanouar 1995). The global changes that the planet has experienced make the conservation of biodiversity, in its entirety, a global issue. The incomplete state of our knowledge on the inventory of species present on the scale of Morocco and their modes of distribution severely limits the proposal of an adequate policy for the conservation of biodiversity.

CONCLUSION

The assessment made during this work clearly shows that the region of Tafilalt has an important specific richness which is also undervalued taxonomically and unfortunately subject to several threats affecting its richness, its areas of distribution and consequently the increase of its endemism see its disappearance.

According to the distribution maps developed, each taxon follows a particular distribution. In addition, Amphipods, Isopods and Gastropods cover large areas of distribution.

The spatial database is an essential tool for identifying regions of high biodiversity (species richness and endemism) and priority areas for conservation. Continued environmental degradation affects the stability of underground aquatic habitats. In addition, ongoing, extremely rapid, human-induced global changes (climate change, drought, floods, deforestation, habitat destruction, and the spread of alien and invasive species) continue to deplete biological diversity.

The global changes that the planet has experienced make the conservation of biodiversity, in its entirety, a global issue. The incomplete state of our knowledge on the inventory of species present on the scale of Morocco and their modes of distribution severely limits the proposal of an adequate policy for the conservation of biodiversity. A few sites have been identified. The sites containing a particular stygological wealth are called "hotspots". In such a conservation strategy, it is the "hot spots" that have priority for preservation.

REFERENCES

Alroy, J. (2002). How many named species are valid? *Proceedings of the National Academy of Sciences of the United States of America, 99*(6), 3706–3711. doi:10.1073/pnas.062691099 PMID:11891342

Alroy, J. (2003). Taxonomic inflation and body mass distributions in North American fossil mammals. *Mammalogy, 84*(2), 431–443. doi:10.1644/1545-1542(2003)084<0431:TIABMD>2.0.CO;2

Ben Azzouz, A. (1983). *Recherche de la faune interstitielle dans la vallée de l'oued Gheris (versant Sud du Haut-Atlas Occidental)* [Search for interstitial fauna in the valley of the Wadi Gheris (southern slope of the Western High Atlas).]. Mém. de CEA, Univ. Cadi Ayyad, Fac. Sc. Semlalia.

Boughrous, A. A. (2007). Biodiversité, Ecologie et Qualité des eaux souterraines de deux régions arides du Maroc: le Tafilalet et la région de Marrakech. [Biodiversity, Ecology and Groundwater Quality of two arid regions of Morocco: Tafilalet and the Marrakech region.] [Thesis, University Cadi Ayyad, North Africa].

Boughrous, A. A., Boulanouar, M., Yacoubi-Khebiza, M., & Coineau, N. (2007). The first *Microcharon* (Crustacea, Isopoda, Microparasellidae) from the Moroccan north Saharan platform. Phylogeny, origin and paleogeography. *Contributions to Zoology (Amsterdam, Netherlands), 76*(1), 19–32.

Boulanouar, M. (1982). Premieres recherches sur la faune des puits de Marrakech, Mé thodes de capture de la faune stygobie par piégeage. [First research on the fauna of the wells of Marrakech, Methods of capture of the stygobic fauna by trapping.] *Mémoire de DEA de 3e cycle, Faculté des Sciences de Marrakech, 41.*

Boulanouar, M. (1995). Faune aquatique des puits et qualité de l“eau dans la région de Marrakech et des Jbilet. Statut et dynamique d“une population de *Proasellus coxalis africanus* (Crustacés: Isopodes) des Jbilet. [Aquatic fauna of wells and water quality in the region of Marrakech and Jbilet. Status and dynamics of a population of *Proasellus coxalis africanus* (Crustaceans: Isopods) of Jbilet.] [Thesis, University Cadi Ayyad, North Africa].

Boutin, C. (1993a). Biogéographie historique des Crustacés Isopodes Cirolanides stygobies du groupe Typhlocirolana dans le bassin méditerranéen. [Historical biogeography of Isopod crustaceans Cirolanides stygobies of the Typhlocirolana group in the Mediterranean basin.] Comptes Rendus de l'Académie des Sciences Paris. *Sciences de la Vie, 316*(12), 1505–1510.

Boutin, C. (1993b). Biogéographie historique des Crustacés Malacostracés stygobies du Maroc. [Historical biogeography of the stygobic Malacostraceous crustaceans of Morocco.] [Thesis, University Claude Bernard, France].

Boutin, C. (1993c). Stygobiology and Historical Geology: The age of the Fuerteventura Island (Canary) inferred from its present stygofaune. *Bull. Soc. Géol, 6*, 220-245.

Boutin, C., Boulal, M., Boulanouar, M., Coineau, N., Ghlala, A., Merzoug, D., ... & Yacoubi-Khebiza, M. (2011). Importance, dans les zones arides et semi-arides, de la biodiversité des faunes aquatiques souterraines. [Importance of biodiversity of underground aquatic fauna in arid and semi-arid areas.] *Actes de Séminaire International sur la Biodiversité Faunistiques en Zones Arides et Semi-arides*, (pp. 132-149).

Boutin, C., & Coineau, N. (1990). "Regression Model", "Modèle biphase d'évolution" et origine des micro-organismes stygobies interstitiels continentaux. *Revue de Micropaleontologie*, *33*, 303–322.

Chappuis, P. A. (1953). Un nouvel isopode psammique du Maroc: Microcerberus remyi. [A new psammic isopod from Morocco: Microcerberus remyi.]. *Vie et Milieu*, *4*(4), 659–663.

Delamare Deboutteville, C.D. (1960). Biologie des eaux souterraines littorales et continentales. [Biology of coastal and continental groundwater.]

Dray, S. (2003). Eléments d''interface entre analyses multivariées, systèmes d''informations géographiques et observations écologiques. [Interface elements between multivariate analyses, geographic information systems and ecological observations.] [Thesis, University Claude Bernard, France].

Driouech, F., Déqué, M., & Mokssit, A. (2009). Numerical simulation of the probability distribution function of precipitation over Morocco. *Climate Dynamics*, *32*(7), 1055–1063. doi:10.100700382-008-0430-6

Gibert, J., Danielopol, D. L., & Stanford, J. A. (1994). Groundwater ecology (Eds). Academic press.

Humphreys, W. F. (2000). Relict faunas and their derivation. In Wilkens H. Culver D.C. and Humphreys W. F. (Eds.) Subterranean ecosystem. Ecosystems of the world 30, (pp. 417- 432). Elsevier.

Khomsi, K., Mahe, G., Tramblay, Y., Sinan, M., & Snoussi, M. (2015). Trends in rainfall and temperature extremes in Morocco. *Natural Hazards & Earth System Sciences Discussions, 3*(2).

Koh, L. P., Dunn, R. R., Sodhi, N. S., Colwell, R. K., Proctor, H. C., & Smith, V. S. (2004). Species coextinctions and the biodiversity crisis. *Science*, *305*(5690), 1632-1634.

Meier, R. J., & Dikow, T. (2004). Significance of specimen databases from taxonomic revisions for estimating and mapping the global species diversity of invertebrates and repatrianting reliable specimen data. *Conservation Biology*, *18*(2), 478–488. doi:10.1111/j.1523-1739.2004.00233.x

Messouli, M. (1994). Évolution, Phylogénie et Biogéographie historique des Metacrangonyctidae, Crustacés Amphipodes stygobies du Nord de l"Afrique et des régions voisines. Thèse d"État. [Thesis, University Cadi Ayyad, North Africa].

Messouli, M., Ben Salem, A., Ghallabi, B., Yacoubi-Khebiza, M., Boughrous, A. A., El Alami El Filali, A., & Ezzahra Hammadi, F. (2009). Ecohydrology and groundwater resources management under global change: A pilot study in the pre-Saharan basins of southern Morocco. *Options Méditerranéennes*, *88*, 255–264.

Pimm, S. L. & Lawton, J.H. (1998). Planning for biodiversity. *Science*, *279*, 2068-2069.

Sihvonen, P., & Siljander, M. (2005). Species diversity and geographical distribution of Scopulini moths (Lepidoptera: Geometridae, Sterrhinae) on world-wide scale. *Biodiversity and Conservation*, *14*(3), 703–721. doi:10.100710531-004-3921-8

Silverman, B. W. (1986). *Density Estimation for Statistics and Data Analysis*. Chapman and Hall.

Soberon, J. M., Llorente, J. B., & Onate, L. (2000). The use of specimen- label databases of conservation purposes: An example using Mexican Papilionid and Pierid butterflies. *Biodiversity (Nepean)*, *9*, 1441–1466.

Stork, N. E., & Samways, M. J. (1995). Inventorying and monitoring biodiversity. In V. H. Heywood & R. T. Watson (Eds.), *Global biodiversity assessment*. Univ. Press.

Vitousek, P. M., Mooney, H. A., Lubchenco, J., & Melillo, J. M. (1997). Human domination of earth"s ecosystem. *Science*, *277*(5325), 294–499. doi:10.1126cience.277.5325.494

ADDITIONAL READING

KHALDOUN, L., MERZOUG, D., & BOUTIN, C. (2013). Faune aquatique et qualité de l'eau des puits et sources de la région de Khenchela (Aurès, Algérie nord-orientale). [Aquatic fauna and water quality of wells and springs in the Khenchela region (Aurès, north-eastern Algeria).]. *Bulletin de la Société Zoologique de France*, *138*(1-4), 273–292.

Steli, S., Hilali, M., Mahboub, A., Rachid, L., & Barbot, A. (2019). Groundwater quality assessment of the Jorf-Fezna palm grove (South-Eastern Morocco). *EWASH & TI Journal*, *3*(4), 231–237.

Chapter 5
The Sustainability of Seafood Market Under the Attack of Invasive Alien Species:
A Review of Turkish Marines

Merve Kaplan
https://orcid.org/0000-0001-5096-5546
Canakkale Onsekiz Mart University, Turkey

Seda Yildirim
https://orcid.org/0000-0003-4367-6652
Tekirdag Namık Kemal University, Turkey

ABSTRACT

This study aims to review the risk of invasive alien species in Turkish marines in the context of sustainable seafood market. Thus, this study investigates the challenges of invasive alien species on seafood market and marine biodiversity. By using classifications, tables, and figures, this study will determine the current profile of Turkish seafood market due to the risk level with alien species. The open access data obtained from Turkish Statistical Institute (TURKSTAT) and FAO (Food and Agriculture Organization of the United Nations) will be used to give main indicators for Turkish seafood market. Then, recent reports and news will be used to determine the efficiency of the implications and policies to fight against to alien species in Turkey. By reviewing recent news and reports, it is thought to give a holistic approach for the related literature.

DOI: 10.4018/978-1-7998-9289-2.ch005

INTRODUCTION

The recent literature has shown that food security and water security were both vital goals to be achieved in the long term. In other words, food security and water security are related together to design sustainable future. Water sources (freshwater, clean and drinking) support the sustainability of food sources such as agriculture and aquaculture. In addition, freshwater sources are vital need for the sustainability of biodiversity and the humanity. There are many challenges in achieving water security such as global warming, climate change, environmental pollution, over-consumption etc. (Bostancı and Yıldırım, 2022; Yildirim et al., 2022; Yıldırım et al., 2018). Keeping the higher quantity and quality of wetlands seems to be so hard while the demand of water is rising. On the other side, wetlands provide alternative food sources for many countries by aquaculture and fishing. At this point, it can be said that policy makers should protect marine wetlands to support water security and food security in the long term.

The impact of climate change and invasive species are considered to be important variables that lead to the destruction of the local ecosystem, especially in the seas. Although the depletion of food resources and local biodiversity are the leading effects of invasive species, it is a fact that the seas and countries are also affected economically (Occhipinti-Ambrogi, 2007). Both scientists and policy makers say that the introduction of alien species should be seen as the biggest threat that countries will face in the future. When a increase in the introduction of foreign species, there will be a decrease in the population of local biodiversity and food sources (Bax et al., 2003). Sustainable seafood market is seen as an important food market as supporting food security in general. Especially, seafood market is primary food source for developing countries and least developing countries in the world. On the other side, many developed countries benefits from seafood to support society's health in the long term (Yıldırım and Yıldırım, 2021; Yıldırım and Kaplan, 2020). Turkey which produces 40 percent of the global supply in the global fish producers market, took its place in the ranking as the ninth country due to its higher production of sea bass and sea bream. Rainbow trout has the largest share in freshwater fish production. However, the country's exports of fish, crustaceans, molluscs and other aquatic invertebrates have continued to increase in the recent period, according to data from TURKSTAT. In addition, bne IntelliNews (*news from Eastern Europe, Eurasia, the Middle East and Africa*) reported that Turkish aquaculture exports were reported as *\$962.23 million in 2019 and over \$1 billion in 2020* (The Fish Site, 2021). Seafood market is an important industry for the European region. However, the growth of EU seafood production did not show great performance in 2018 and 2019. To sustain sustainable seafood market including low priced seafood with enough quantity, is almost related to sustain seafood import (European Union, 2019). According to

CBI (*What is the demand for fish and seafood on the European market?,* 2021), Europe is a very good market from the point of view of seafood exporters. Although the European seafood market has been affected by the Covid-19 crisis, a certain steady trend in seafood demand has been caught in 2020. No significant decrease or shrinkage was observed in seafood imports in Europe during the Covid-19 crisis. In fact, it can be said that Europe has the title of the second largest seafood importer in the world. Europe imported $19.8 billion of seafood from non-European regions in 2020. At this point CBI (*What is the demand for fish and seafood on the European market?,* 2021):

- China, with exporting fish fillet ($1.0 billion),
- Vietnam with exporting of pangasius and shrimp ($821 million),
- Ecuador with exporting shrimp and tuna export ($490 million),
- India with exporting shrimp ($361 million),
- Turkey with exporting sea bass and sea bream and fish fillets ($315 million).

During the last decades, it has been observed that there are significant declines in the quantity of captured marine fish reserves in the Asia and Europe region. When considering the importance of blue economy, new measures and policies to achieve sustainable seafood market is a vital issue in the long term. As given in table 1, the quantity of captured marine fish reserves from Asia and Europe has decreased in last decades.

Table 1. The quantity of captured marine fishes in Aegean, Black Sea and Marmara Sea (tonnes)

Land Areas	Ocean Area	2011	2012	2013	2014	2015	2016	2017	2018
Asia	Aegean	27.414	30.999	27.764	30.056	30.788	29.998	41.474	36.951
	Black Sea	352.558	191.534	213.552	161.632	293.476	182.161	285.300	192.205
	Marmara Sea	35.317	42.187	37.043	29.991	28.568	27.377	20.930	48.565
Total Asia		415.289	264.730	278.359	221.679	352.832	239.536	347.704	277.722
Europe	Aegean	41.578	41.656	41.817	40.779	44.763	50.807	53.604	52.705
	Black Sea	69.655	64.100	71.924	41.262	81.616	95.442	75.927	61.513
Total Europe		111.233	105.756	113.742	82.040	126.378	146.230	129.532	114.218
Grand total		526.522	370.486	392.101	303.719	479.210	385.766	477.236	391.939

Source: adapted from FAO - Fisheries and Aquaculture Information and Statistics Branch (2021)

Climate change due to global warming has revealed the danger in the seas. Increasing water temperature causes changes in migration routes and an increase in invasive species. It is emphasized that this situation will reduce the productivity in fisheries. Scientific views put forward in the workshop on "Climate Change and Our Seas" organized by the Turkish Marine Research Foundation (TÜDAV) and the Middle East Technical University (METU) Marine Sciences Institute in 2020 were collected in the book titled "Climate Change and Its Effects on the Seas of Turkey". The book, which includes the results of the studies on the relationship between the sea change and the marine ecosystem, revealed the danger in the seas due to global warming. According to the book, which states that the Mediterranean has warmed by 1.5 degrees in the last 50 years, an increase in water temperature will increase invasive species, which will reduce the yield of fish (*Danger of invasive species in the seas*, 2021). Marine invasive species are new challenge for countries in the way of achieving sustainable seafood market. For example, live seafood trade increase the population of invasive alien species in marines and threats other seafood products (IUCN (International Union for Conservation of Nature, n.d.). Invasive species are accepted as an important threat to marine biodiversity by NOAA Fisheries. İnvasive marine species can harm marine biodiversity and so harm natural resources in the long term. Invasive marine species can decrease the population of natural species as causing extinction in the related marine. A decrease in marine biodiversity will cause lower stock in seafood sources (NOAA Fisheries (National Oceanic and Atmospheric Administration, n.d.). Accordingly, it can be said that invasive alien species threats economic, social and environmental sustainability. Canada has been troubled with aquatic invasive species in its freshwaters for a long time. But, fight against invasive alien species is not easy in the long term (Government of Canada, n.d.).

This study focuses on reviewing the risk levels of Turkish marine wetlands due to achieving sustainable aquaculture market while challenging with invasive alien species. Seafood products are thought to play an important role in food security for many countries such as Turkey. However, fighting against to invasive alien species can be more difficult to be solved rather than other challenges. For example, legal rules and practices regarding environmental pollution and environmental arrangements can be made. In society, citizens, businesses and local governments can take responsibility and work together for water security and the protection of wetlands. However, different practices or approaches are emerging in dealing with invasive alien species which are not normally biodiversity creatures within these wetlands.

SEAFOOD PRODUCTION IN TURKISH MARINES

Turkey has *783.560* km^2 landing area and *8.333* km of coastline that including "*the Black Sea, the Aegean Sea, the Mediterranean Sea, the Sea of Marmora, the Bosporus and Dardanelles*" Turkey is thoughts to be rich country with having many sources of sea and freshwater food. There are some risks and threats for fishing and seafood production. Although aquaculture production has been increasing rapidly, captured production does not have a stable trend over the years (FAO, n.d.a).

Based on "General Fisheries Commission for the Mediterranean – GFCM", authorized vessels can be presented by Table 2. As seen in Table 2, Turkey is the

Table 2. Authorized vessels among GFCM

Countries	Number of vessels	Percent of vessels (%)	LOA (metres) average of vessels
Egypt	1832	22	18.61
Italy	1432	17	21.20
Turkey	1039	12	26.15
Tunisia	958	12	21.71
Algeria	870	10	19.42
Spain	710	9	21.09
Greece	494	6	22.40
Croatia	307	4	22.41
Morocco	200	2	20.80
Albania	183	2	21.36
France	74	1	26.26
Bulgaria	56	1	19.64
Malta	49	1	21.73
Georgia	35	0	30.45

Source: FAO, (n.d.b)

Figure 1. Fishery Production Trends in Turkey (2008-2020)
Source: TURKSTAT, 2021a

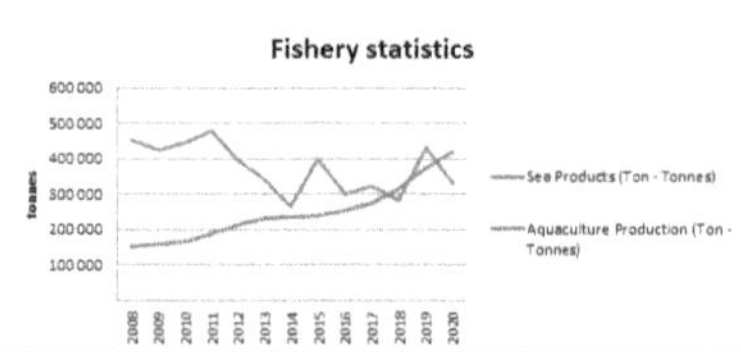

Table 3. Major species groups in capture production for Turkey (tonnes)

Year	Pelagic Marine Fish	Demersal Marine Fish	Freshwater and Diadromous Fish	Molluscs excl. Cephalopods	Crustaceans	Cephalopods	Aquatic Animals NEI	Marine Fish NEI
1980	354875	36886	25048	1269	6533	355	4	1875
1981	395772	39012	24302	2200	7521	518	44	787
1982	425615	39847	25183	4312	7921	176	0	403
1983	455047	56002	30033	2341	11843	480	0	1541
1984	467365	42087	36452	2971	14735	887	1959	474
1985	474449	46532	35343	4602	14217	830	1719	377
1986	477047	49086	34468	8820	6288	986	2802	344
1987	501094	61801	36556	15517	8132	1295	3	296
1988	507572	70688	42597	23899	10622	9730	1821	2971
1989	288138	71397	40172	27289	10196	11672	1173	448
1990	230517	62171	35695	30809	7581	8077	1109	3241
1991	210001	78229	37490	25908	2102	952	1321	946
1992	283210	83785	39107	33944	3660	1747	1212	708
1993	365276	85099	42252	43583	4800	1395	1531	1596
1994	401355	88895	42799	45429	4064	1955	1665	1412
1995	481221	78128	41762	22139	2681	1866	1351	1829
1996	363247	91226	38764	19583	2299	1810	1644	1855
1997	308113	79295	45105	19282	2883	2320	1060	1095
1998	316389	97970	49580	14107	3243	2700	1850	1361
1999	425866	85352	45588	11524	2457	1407	1321	309
2000	349143	86743	37913	15680	3894	1630	977	7365
2001	394495	72160	37510	13435	4935	2095	2873	230
2002	449427	45219	37679	23751	6137	2741	1398	330
2003	377956	39151	38409	36604	8392	1985	4792	483
2004	419853	38237	38831	40279	7801	3261	1803	417
2005	295241	42238	38728	38112	7351	2808	1347	671
2006	372891	41759	35103	72116	4862	3285	1850	1182
2007	478921	43684	34998	65557	4788	2796	895	811
2008	365737	34885	33189	50977	5546	2720	668	396
2009	347310	36512	31520	39379	5509	2483	622	582
2010	361662	40657	33082	40575	5836	2634	780	699
2011	407177	29817	28879	40100	5456	1879	750	705
2012	289375	29090	30740	74505	5577	2288	648	221
2013	272573	26976	27716	39231	4592	2019	831	190
2014	211054	25145	27694	30781	5008	1361	742	429
2015	321698	27958	28320	47363	4540	1327	535	168
2016	240574	27016	27375	32986	5056	1560	486	273
2017	249636	24466	25033	47333	5418	1571	547	316
2018	203077	22824	23494	57093	5094	1790	487	236
2019	353346	24878	24737	51670	5847	1853	454	383

Source: FAO Fisheries statistics, (n.d.)

Table 4. The quantity of caught sea fish (tonnes)

Type of fish	Latin names	2011	2020
Leer fish	Lichia amia	585.9	141.7
Greater amberjack	Seriola dumerili	31.4	9.7
Albacore	Thunnus alalunga	1 395.7	16.2
Hake-European hake	Merluccius merluccius	921.1	1 148.7
Red mullet	Mullus surmuletus	1 861.4	1 604.2
Goldon banded	Gnathanodon speciosus	427.8	35.5
Sprat	Sprattus sprattus	87 140.8	26 803.8
Sea bream	Sparus aurata	766.1	583.7
Common sole	Solea solea	829.3	457.5
John dory	Zeus faber	67.4	50.0
Common sea bream	Pagrus pagrus	69.7	35.4
Angler fish	Lophiiformes	193.0	208.2
Shore rockling	Gaidropsarus mediterraneus	15.3	37.2
Frigate mackerel	Auxis thazard	2 551.8	1 069.8
Meagre	Argyrosomus regius	30.9	67.2
Anchovy	Engraulidae	228 491.4	171 253.2
Painted comber	Serranus scriba	34.2	11.6
European barracuda	Sphyraena sphyraena	228.1	94.0
Black scorpion fish	Scorpaena porcus	196.4	107.7
Annular bream	Diplodus annularis	195.9	54.0
Horse mackerel	Trachurus trachurus	18 072.7	7 494.8
Scad	Trachurus trachurus	6 937.3	4 854.5
Brown meagre	Sciaena umbra	6.6	0.8
Picarel	Esox Lucius	877.5	178.8
Turbot	Scophthalmus maximus	166.4	412.0
Two banded bream	Diplodus vulgaris	152.6	78.9
Gobies	Gobiidae	95.8	31.8
Grey mullet	Mugil cephalus	2 513.8	1 415.5
Angel shark	Squatina squatina	15.7	0.2
Sword fish	Xiphias gladius	189.6	402.4
Red gurnard	Chelidonichthys cuculus	211.6	28.6
Trigla lineata	Chelidonichthys lastoviza	54.7	2.9
Chup mackerel	Scomber scombrus	3 127.0	2 239.2
Topeshark	Galeorhinus galeus	369.5	4.7
Bogue	Boops boops	2 113.5	2 598.8
Waker	Epinephelus aeneus	396.6	248.1
Sea bass	Dicentrarchus labrax	316.5	134.7
Small-Scalled	Scorpaena porcus	84.2	12.6
Blue fish	Pomatomus saltatrix	3 122.0	3 722.4
Saddled sea bream	Oblada melanura	112.8	64.6
Striped bream	Diplodus vulgaris	635.6	948.0
Whiting	Merlangius merlangus	9 454.8	9 363.7
European coger	Conger conger	0.8	0.1
Dusky grouper	Epinephelus marginatus	34.0	0.8
Bluefin tuna	Thunnus thynnus	527.5	2 252.0
Little tunny	Euthynnus alletteratus	1 437.4	340.9
Piper	Hyporhamphus ihi	15.2	5.1
Atlantic bonito	Katsuwonus pelamis	10 018.9	22 742.7
Large-eye dentex	Dentex macrophthalmus	53.6	28.3
Flounder	Paralichthys	47.3	5.7
Pilchard	Sardina pilchardus	34 708.6	21 265.4
Black sea bream	Spondyliosoma cantharus	23.8	41.3
Saupe	Sarpa salpa	166.9	95.0
Dentex	Dentex dentex	82.8	70.1
Sharpsnout sea bream	Diplodus puntazzo	13.8	1.1
Striped red	Sciaenops ocellatus	3 876.5	2 774.8
Twaite shad	Alosa fallax	2 581.5	2 611.7
Mackerel	Scomber scombrus	147.3	172.6
Thornback ray	Raja clavata	401.0	3.9
Gar fish	Lepisosteidae	317.1	366.9
Saury	Cololabis saira	319.2	179.7
Other		673.1	351.7
Total		**432 246.0**	**291 910.1**

Source: adapted from TURKSTAT, 2021b

third country among GFCM with having 12 percent vessels in total. Accordingly, it can be said that Turkey plays an important role in this region.

Fishing and aquaculture is an important sector in Turkey. Considering its economic, social and environmental value, the problem of protection and sustainability of freshwater areas requires that it be addressed within the scope of both food security and water security. The importance of wetlands in this study provides a closer look at the food security field. In this study, the threat of alien species to the safety of seafood is examined. For the purpose of the study, it will be useful to give a profiling of the products obtained from the wetlands in Turkey. At this point, the current profile for the seafood market in Turkey can be given as Table 3:

Table 5. The quantity of caught other seafood products (tonnes)

Type of seafood products	Latin names	2011	2020
Octopus	Octopoda	321.8	310.7
Spiny lobster	Palinuridae	25.8	0.5
Norway lobster	Nephrops norvegicus	24.8	6.9
Sea snail	Semicassis granulate	6 533.8	8460.7
Common lobster	Nephropidae	4.7	2.7
Oystre	Ostrea	5.9	-
Long finned squid	Loligo forbesii	394.1	631.4
Speckled shrimp	Metapenaeus monoceros	301.2	26.6
Green tiger prawn	Penaeus monodon	543.4	551.9
Caramote prawn	Melicertus kerathurus	642.9	171.6
Giant gamba prawn	Aristaeomorpha foliacea	1 800.9	939.2
Deep water rose prawn	Parapenaeus longirostris	1 481.5	3514.7
Carpet Shell	Ruditapes decussatus	26.7	57.1
Striped venus	Chamelea gallina	30 175.6	21 823.6
Mediterranean mussel	Mytilus galloprovincialis	1 458.8	1 034.8
Bearded horse mussel	Modiolus barbatus	347.2	3.2
Cuttle fish	Sepiida	1 163.3	960.5
Common shore crab	Carcinus maenas	8.7	3.7
Great Scallop	Pecten maximus	17.8	0.3
Blue crab	Callinectes sapidus	10.7	3.4
Sea cucumber	Holothuroidea	-	829.3
Other		122.8	37.9
Total		45 412.4	39 370.7

Source: adapted from TURKSTAT, 2021c

Table 6. Intended use of landings (tonnes) in Turkey (2020)

Type of fish	Latin names	Intended Use of Landings (2020)				
		Quantity of catches	Quantity of landings	Human consumption	Industrial Use	Other Use
Leer fish	Lichia amia	141,7	141,7	141,7	-	-
Greater amberjack	Seriola dumerili	9,7	9,7	9,7	-	-
Albacore	Thunnus alalunga	16,2	16,2	16,2	-	-
Hake-European hake	Merluccius merluccius	1 148,7	1 148,5	1 148,5	-	-
Red mullet	Mullus surmuletus	1 604,2	1 603,0	1 603,0	-	-
Goldon banded	Gnathanodon speciosus	35,5	35,0	35,0	-	-
Sprat	Sprattus sprattus	26 803,8	26 803,8	-	26 803,8	-
Sea bream	Sparus aurata	583,7	583,0	583,0	-	-
Common sole	Solea solea	457,5	457,5	457,5	-	-
John dory	Zeus faber	50,0	50,0	50,0	-	-
Angler fish	Lophiiformes	208,2	208,2	208,2	-	-
Shore rockling	Gaidropsarus mediterraneus	37,2	37,2	37,2	-	-
Frigate mackerel	Auxis thazard	1 069,8	1 069,6	1 069,6	-	-
Meagre	Argyrosomus regius	67,2	67,2	67,2	-	-
Anchovy	Engraulidae	59 373,2	59 365,1	59 342,2	22,9	-
Fish meal-oil factories	-	111 880,0	111 879,5	-	111 879,5	-
Painted comber	Serranus scriba	11,6	11,6	11,6	-	-
European barracuda	Sphyraena sphyraena	94,0	94,0	94,0	-	-
Black scorpion fish	Scorpaena porcus	107,7	107,7	107,7	-	-
Annular bream	Diplodus annularis	54,0	54,0	54,0	-	-
Horse mackerel	Trachurus trachurus	7 494,8	7 490,2	7 490,2	-	-
Scad	Trachurus trachurus	4 854,5	4 851,6	4 851,6	-	-
Brown meagre	Sciaena umbra	0,8	0,8	0,8	-	-
Picarel	Esox Lucius	178,8	178,8	178,8	-	-
Turbot	Scophthalmus maximus	412,0	412,0	412,0	-	-
Two banded bream	Diplodus vulgaris	78,9	78,9	78,9	-	-
Gobies	Gobiidae	31,8	31,8	31,8	-	-
Grey mullet	Mugil cephalus	1 415,5	1 415,5	1 415,5	-	-
Angel shark	Squatina squatina	0,2	0,2	0,2	-	-
Sword fish	Xiphias gladius	402,4	402,4	402,4	-	-
Red gurnard	Chelidonichthys cuculus	28,6	28,6	28,6	-	-
Trigla lineata	Chelidonichthys lastoviza	2,9	2,9	2,9	-	-
Chup mackerel	Scomber scombrus	2 239,2	2 239,2	2 239,2	-	-
Topeshark	Galeorhinus galeus	4,7	4,7	4,7	-	-
Bogue	Boops boops	2 598,8	2 598,8	2 598,8	-	-
Waker	Epinephelus aeneus	248,1	248,1	248,1	-	-
Sea bass	Dicentrarchus labrax	134,7	134,7	134,7	-	-
Small-Scalled	Scorpaena porcus	12,6	12,6	12,6	-	-
Blue fish	Pomatomus saltatrix	3 722,4	3 722,1	3 722,1	-	-
Saddled sea bream	Oblada melanura	64,6	64,6	64,6	-	-
Striped bream	Diplodus vulgaris	948,0	948,0	948,0	-	-
Whiting	Merlangius merlangus	9 363,7	9 300,9	9 300,9	-	-
Dusky grouper	Epinephelus marginatus	0,8	0,8	0,8	-	-
Bluefin tuna	Thunnus thynnus	2 252,0	2 252,0	2 252,0	-	-
Little tunny	Euthynnus alletteratus	340,9	340,9	340,9	-	-
Piper	Hyporhamphus ihi	5,1	5,1	5,1	-	-
Atlantic bonito	Katsuwonus pelamis	22 742,7	22 740,3	22 740,3	-	-
Large-eye dentex	Dentex macrophthalmus	28,3	28,3	28,3	-	-
Flounder	Paralichthys	5,7	5,7	5,7	-	-

continued on following page

Table 6. Continued

Type of fish	Latin names	Intended Use of Landings (2020)				
		Quantity of catches	Quantity of landings	Human consumption	Industrial Use	Other Use
Pilchard	Sardina pilchardus	21 265,4	21 265,4	21 265,4	-	-
Black sea bream	Spondyliosoma cantharus	41,3	41,3	41,3	-	-
Saupe	Sarpa salpa	95,0	95,0	95,0	-	-
Dentex	Dentex dentex	70,1	70,1	70,1	-	-
Sharpsnout sea bream	Diplodus puntazzo	1,1	1,1	1,1	-	-
Striped red	Sciaenops ocellatus	2 774,8	2 767,4	2 767,4	-	-
Twaite shad	Alosa fallax	2 611,7	2 610,7	2 610,7	-	-
Pink dentex	Dentex gibbosus	26,2	26,2	26,2	-	-
Mackerel	Scomber scombrus	172,6	172,6	172,6	-	-
Thornback ray	Raja clavata	3,9	3,9	3,9	-	-
Gar fish	Lepisosteidae	366,9	366,9	366,9	-	-
Saury	Cololabis saira	179,7	179,7	179,7	-	-
Other		351,7	351,7	351,7	-	-
Total		**291 910,1**	**291 817,3**	**153 111,1**	**138 706,2**	**0,0**

Source: TURKSTAT, 2021d

Figure 1 shows the trend of fishery production in Turkey. The recent years has shown that the quantity of caught fish production was decreasing while the quantity of aquaculture production was increasing.

Table 4 shows the quantity of caught fish from Turkish marines for the year of 2011 and 2020. When comparing the amount of fish caught in 2011 and the amount of fish caught in 2020, it is seen that there has been a significant decrease in the most of fish species recently.

Table 5 shows the quantity of caught other seafood products in Turkish marines. It can be said that the amount of other seafood caught in 2011 did not decrease significantly in 2020. According to this, it can be said that the amount of other sea products compared to sea fish has not experienced a great decrease in the last ten years.

Table 6 shows fish species and the distribution of these fish species according to usage type by 2020.

Table 7 shows other seafood species and the distribution of these fish species according to usage type by 2020.

Turbot fishery in Black Sea is an important area that main information is given as Table 8:

Based on FAO, table 9 shows main info for vessels in the Levant Sea.

Table 7. Landing of other sea products for intended use, 2020 (tonnes)

Other sea products	Latin Names	Intended Use of Landings				
		Quantity of catches	Quantity of landings	Human consumption	Industrial Use	Other Use
Octopus	Octopoda	310,7	310,7	310,7	-	-
Carpet shell	Ruditapes decussatus	57,1	57,1	57,1	-	-
Striped venus	Chamelea gallina	21 823,6	21 823,6	21 823,6	-	-
Mediterranean mussel	Mytilus galloprovincialis	1 034,8	1 034,8	1 034,8	-	-
Bearded horse mussel	Modiolus barbatus	3,2	3,2	3,2	-	-
Warty venus	Venus verrucosa	-	-	-	-	-
Spiny lobster	Palinuridae	0,5	0,5	0,5	-	-
Norway lobster	Nephrops norvegicus	6,9	6,9	6,9	-	-
Sea snail	Semicassis granulate	8 460,7	8 460,7	8 460,7	-	-
Common lobster	Nephropidae	2,7	2,7	2,7	-	-
Oystre	Ostrea	-	-	-	-	-
Squid	Decabrachia	631,4	631,4	631,4	-	-
Green tiger prawn	Penaeus monodon	551,9	551,9	551,9	-	-
Caramote prawn	Melicertus kerathurus	171,6	171,6	171,6	-	-
Giant gamba prawn	Aristaeomorpha foliacea	939,2	939,2	939,2	-	-
Deep water rose prawn	Parapenaeus longirostris	3 514,7	3 514,5	3 514,5	-	-
Speckled shrimp	Metapenaeus monoceros	26,6	26,6	26,6		-
Cuttle fish	Sepiida	960,5	960,5	960,5	-	-
Common shore crab	Carcinus maenas	3,7	3,7	3,7	-	-
Great Scallop	Pecten maximus	0,3	0,3	0,3	-	-
Blue crab	Callinectes sapidus	3,4	3,4	3,4	-	-
Sea cucumber	Holothuroidea	829,3	829,3	829,3	-	-
Other		37,9	37,9	35,0	2,6	0,3
Total		39 370,7	39 370,5	39 367,6	2,6	0,3

Source: TURKSTAT, 2021e

Table 8. Some selected info about vessels in turbot fishery

Countries	Number of authorized vessels by CPC	LOA average (metres) of authorized vessels by CPC	Capacity average (GT) of authorized vessels by CPC
Turkey	1172	7.31	3.53
Ukraine	461	7.67	1.23
Bulgaria	126	13.80	21.92
Romania	59	10.22	13.99

Source: FAO, (n.d.c).

Table 9. Selected info for vessels to red shrimp and blue and red shrimp

Countries	Number of authorized vessels	LOA average (metres) of vessels	Engine power average (kW)	Capacity average (GT)
Turkey	78	21.81	383.57	63.17
Italy	34	28.29	451.35	163.00
Israel	16	18.16	215.68	62.30
Cyprus	5	22.73	334.97	107.21

Source: FAO, (n.d.d).

THE RISK OF INVASIVE ALIEN SPECIES IN TURKISH MARINES

Invasive alien species are considered as species that have settled in areas other than natural areas and pose a threat to local biodiversity and ecosystems due to their presence and spread in these areas. In general, there is no ecosystem in the world that can withstand alien invasive species. The success of species introduced or settling in a new environment often depends on a combination of several bio-ecological factors. However, some ecosystems are known to be more susceptible to invasion, as has been largely experienced in the Mediterranean. The influx of large numbers of tropical species into the Mediterranean is undoubtedly the most remarkable biogeographic phenomenon of our time. Invasive species can overwhelm local species, causing the balance of the ecosystem to reach an irreversible point. After the opening of the Suez Canal in 1869, due to the removal of the salt barriers due to the canal widening and deepening works, the Red Sea species came to the Mediterranean and the two ecosystems met. The characteristics of the Mediterranean water have allowed species from the Red Sea to colonize over the years (WWF, 2018).

Table 10. A recent news on the risk of invasive alien species in Turkish Marines

News source	Interview with	Result
Invasive 2 Species Take Place Among Mediterranean Fish (2020)	Prof. Dr. M. Gökoğlu (from Akdeniz University)	Prof. Dr. Mehmet Gökoğlu (from Akdeniz University) said that these two species were listed in the fish inventory of the Mediterranean and Marine Biology Museum in Antalya, Turkey. Prof. Gökoğlu determined that there were sub-species of Puffer Fish and Lion Fish that have been migrated to the Mediterranean recently. He said that the poisonous Stonefish-Synanceia verrucosa also lived in the Mediterranean. Prof. Gökoğlu pointed out that the Mediterranean ecosystem was now changing and native species started to be replaced by Red Sea species (*Invasive 2 Species Take Place Among Mediterranean Fish*, 2020).
Akgüneş, (2020)	Prof. Dr. Köşker (from Çukurova University)	According to Prof. Dr. Köşker, it was no longer possible to get rid of this species in the Mediterranean, because the pufferfish had no predators in the ecosystem (Akgüneş, 2020).
Özacar, (2020).	Assoc. Prof. Topaloglu (from Istanbul University)	Assoc. Prof. Topaloglu (from Istanbul University) determined that it was not possible for alien species to create a habitat for themselves without destroying native species. Therefore, it was seen that the number of foreign sea urchins increased instead of domestic sea urchins. In the long term, an uncertainty and imbalance in marine biological life awaits humanity (Özacar, 2020).
DHA (Demirören News Agency) (2018)	The international joint press release of the Turkish Marine Research Foundation (TUDAV) on 'Toxic Alien Species'	It was noted that a small portion of the alien invasive species have dangerous effects on human health due to their poisonous nature (DHA, 2018): a) Lionfish (Pterois miles): This fish can be eaten with care. Yet their presence on the coastline, popular with fishermen and tourists, poses a threat to human health. b) Sokar Fish (Siganus rivulatus, Siganus luridus):These edible species are found in large flocks in the Southeast Mediterranean and are now expanding both north and west. Fishermen should be careful when hunting and cleaning these fish. The meat of both species is edible when handled and cleaned carefully. c)Migratory Jellyfish (Rhopilema nomadica): It is dangerous species.
Çalkaya, (2020)	Ramazan Özkaya (the Chairman of the Board of Directors of the Central Union of Fisheries Cooperatives)	Mr. Özkaya said that puffer fish harms both the ecosystem and the fishing industry. According to Özkaya, the reduction of the numbers of puffer fish is very important for the sustainability seafood market (Çalkaya, 2020).
Özacar (2019)	Prof. Dr. F.S. Karakulak (from Istanbul University)	Prof. Dr. F.S. warned policy makers about the rapid decline in fisher stock and production in Turkish marines. Climate change, global warming, overfishing threatened the sustainability of seafood products.
Sarı (2021)	Prof. Dr. H. Filiz (from Mugla Sıtkı Kocman University)	Invasive marine species threatened local marine biodiversity in the Black Sea.

Source: created by authors

In the last 10 years, there has been a 30 percent decline in the total amount of fish caught by Turkey from its seas. The most striking decrease was seen in bottom fish species such as mullet, sea bass, sea bream, leer, and turbot. According to the TUIK (2014) report, the amount of fish obtained from the seas of Turkey decreased by 30 percent from 334 thousand 248 tons in 2005 to 231 thousand tons last year. Anchovy, sardine and horse mackerel, which were among the most hunted species, experienced the worst season of the last 10 years. While 96 thousand 440 tons of anchovy, 18 thousand 77 tons of sardines and 12 thousand 213 horse mackerel were caught last year, the most striking decreases were seen in bottom fish. Dr. Ali Cemal Güclü (METU Marine Sciences Institute Faculty Member) stated the population of many species was expected to continue declining due to increasing sea temperature in Turkey (Albay, 2015). Prof. Dr. Firdes Saadet Karakulak (Head of Istanbul University Fisheries Technology and Management Department) emphasized that “overfishing” should be prevented in relation to the decrease in fish stocks. However, marine pollution, global climate change, and other invasive species that come to Turkey’s sea waters with the ballast waters of ships also negatively affect fish stocks. According to Prof. Karakulak, the fish reduce the first breeding length in order to maintain their own lineage, if the stock tends to decrease. In fact, it is possible to increase fish stocks and sustainable fishing by making good fisheries management (Özacar, 2019).

In Turkey, the biggest risk of invasive alien species belongs to the Mediterranean region. There are some news and reports that summarizing the risk level of the Mediterranean Sea and other Turkish marines due to increasing numbers of invasive alien species as table 10:

When reviewing the recent news on invasive alien species in Turkish marines, it is seen that the population of these alien species are decreasing. On the other side, the higher population of invasive alien species decreases the local population of marine biodiversity in Turkish marines. Like as climate change, global warming and overfishing, invasive alien species are also new threats for the sustainability of seafood product.

SOLUTIONS AND RECOMMENDATIONS

The fight against to invasive alien species in Marines is an important issue in the context of 2030 Sustainable Development Goals. To keep sustainable seafood market, the protection of Marine biodiversity and legal measures against over-fishing will be supportive in the long term. On the other side, it is a fact that there is a rising danger coming from alien species in Marines, especially in the Mediterranean. If the population of invasive alien species increases rapidly, local and natural biodiversity will be decrease in the Mediterranean and almost there will be rapid extinction. However, prevent the migration of invasive alien species to the Mediterranean is not so practical due to the climate change. Accordingly, catching of this alien species can be solution to decrease the population of this alien species in the short term.

FUTURE RESEARCH DIRECTIONS

This study reviews the challenges of invasive alien species in Turkish marines in the context of seafood market. The rising number of invasive alien marine species will decrease the population of local species in Turkish marines, especially in the Mediterranean. Climate change and global warming has accelerated the migration of invasive alien species to the Mediterranean. At this point, policy makers will struggle with the higher population of invasive alien species in Turkish marines as keeping sustainable seafood market. Future studies can review how invasive alien species influence the seafood market by empirical analysis or other methods.

CONCLUSION

The literature points out that invasive alien species threats local biodiversity in the Mediterranean Sea (Yildirim and Kaplan, 2022; Katsanevakis et al., 2014, 2016; Mannino and Balistreri, 2021; Galil, 2007, 2008, 2009; Goren et al., 2016; Marchini et.al., 2015; Zenetos et al., 2012; Kourantidou et al., 2021; Servello et al., 2019; Occhipinti-Ambrogi et al., 2021; Marignani et al., 2020).There are several threats to sustainable seafood market including climate change, global warming, environmental pollution, water stress and invasive alien species. Some researchers have revealed the risk condition of sustainable seafood market and marine biodiversity due to the main threats. However, there is a gap in studies that exploring the influence of invasive alien species in the local sea. The mediated displacement of marine species as a result of human activities has led to an unprecedented biodiversity change in the Mediterranean (Çınar et al., 2021; Bilecenoğlu and Çınar, 2021). Therefore, categorizing the number and characteristics of wild alien species entering the Mediterranean according to species groups can provide useful information for the sustainability of the marine market in the future.

This chapter aimed to review recent news and reports or statistics to conclude the risk of invasive alien species in Turkish Marines. Mediterranean Sea has higher quantity of invasive alien species rather than other Turkish marines. The rising temperature of Mediterranean Sea causes climate change and biodiversity in this region. In other words, the Mediterranean Sea has warmer climate recently that welcomes invasive alien species.

Note: Turkish references were translated into English.

REFERENCES

Akgüneş, G. (2020, June 21). *We must learn to live with blowfish.* Milliyet. https://www.milliyet.com.tr/yazarlar/gurkan-akgunes/balon-baligiyla-yasamayi-ogrenmeliyiz-6239709

Albay, T. (2015, July 30). *Fish is declining in Turkey.* AA. https://www.aa.com.tr/tr/turkiye/turkiyede-balik-azaliyor/21379

Bax, N., Williamson, A., Aguero, M., Gonzalez, E., & Geeves, W. (2003). Marine invasive alien species: A threat to global biodiversity. *Marine Policy*, *27*(4), 313–323. doi:10.1016/S0308-597X(03)00041-1

Bilecenoğlu, M., & Çınar, M. E. (2021). Alien Species Threat across Marine Protected Areas of Turkey—An Updated Inventory. *Journal of Marine Science and Engineering*, *9*(10), 1077. doi:10.3390/jmse9101077

Bostancı, S. H., & Yıldırım, S. (2022). The Role of Municipalities in Achieving Water Security: The Case of Turkey. In R. Castanho (Ed.), *Handbook of Research on Sustainable Development Goals, Climate Change, and Digitalization* (pp. 268–286). IGI Global. doi:10.4018/978-1-7998-8482-8.ch017

Çalkaya, M. (2020, December 3). *Fishermen welcomed the promotion of blowfish fishing*. AA. https://www.aa.com.tr/tr/ekonomi/balikcilar-balon-baligi-avciliginin-tesvik-edilmesini-olumlu-karsiladi/2064002

CBI. (2021, August 30). *What is the demand for fish and seafood on the European market?* CBI [the Centre for the Promotion of Imports from developing countries]. https://www.cbi.eu/market-information/fish-seafood/what-demand

Çinar, M. E., Bilecenoğlu, M., Yokeş, M. B., Öztürk, B., Taşkin, E., Bakir, K., Doğan, A., & Açik, Ş. (2021). Current status (as of end of 2020) of marine alien species in Turkey. *PLoS One*, *16*(5), e0251086. doi:10.1371/journal.pone.0251086 PMID:33945562

Denizcilik Dergisi. (2021, April 15). *Danger of invasive species in the seas*. Denizcilik Dergisi [Maritime Magazine]. https://www.denizcilikdergisi.com/denizcilik-gundem-haberler i/denizlerde-istilaci-tur-tehlikesi/

DHA [Demirören News Agency]. (2018, September 28). Attention to 10 species dangerous for human health in the Mediterranean. *CNN Turkey*. https://www.cnnturk.com/turkiye/akdenizde-insan-sagligi-icin -tehlikeli-10-ture-dikkat

European Union. (2019). *The EU Fish Market.* Publications Office of the European Union. https://www.eumofa.eu/documents/20178/314856/EN_The+EU+fish+market_2019.pdf/

FAO. (n.d.a). *Fishery and Aquaculture Country Profiles The Republic of Turkey.* FAO. https://www.fao.org/fishery/facp/TUR/en

FAO. (n.d.b). *GFCM AVL - Authorized Vessel List (>15m LOA).* FAO. https://www.fao.org/gfcm/data/fleet/avl

FAO. (n.d.c). *General Fisheries Commission for the Mediterranean – GFCM.* FAO. https://www.fao.org/gfcm/data/fleet/blackseaturbot

FAO. (n.d.d). *Vessels authorized to fish for red shrimp and blue and red shrimp in the Levant Sea.* FAO. https://www.fao.org/gfcm/data/fleet/levantsea

FAO - Fisheries and Aquaculture Information and Statistics Branch. (2021), *Statistical Query Results, Capture Quantity*. FAO. http://www.fao.org/figis/servlet/SQServlet?file=/usr/local/tomcat/8.5.16/figis/webapps/figis/temp/hqp_4241499935846662994.xml&outtype=html

FAO (n.d.). *Fisheries statistics*. [Data set]. FAO. http://www.fao.org/fishery/facp/TUR/en#CountrySector-Statist ics

Galil, B. S. (2007). Loss or gain? Invasive aliens and biodiversity in the Mediterranean Sea. *Marine Pollution Bulletin*, *55*(7–9), 314–322. doi:10.1016/j.marpolbul.2006.11.008 PMID:17222869

Galil, B. S. (2008). Alien species in the Mediterranean Sea—which, when, where, why? In *Challenges to marine ecosystems* (pp. 105–116). Springer. doi:10.1007/978-1-4020-8808-7_10

Galil, B. S. (2009). Taking stock: Inventory of alien species in the Mediterranean Sea. *Biological Invasions*, *11*(2), 359–372. doi:10.100710530-008-9253-y

Goren, M., Galil, B. S., Diamant, A., Stern, N., & Levitt-Barmats, Y. A. (2016). Invading up the food web? Invasive fish in the southeastern Mediterranean Sea. *Marine Biology*, *163*(8), 1–11. doi:10.100700227-016-2950-7

Government of Canada. (n.d.). *A Canadian Action Plan to Address the Threat of Aquatic Invasive Species*. Government of Canada. https://www.dfo-mpo.gc.ca/species-especes/publications/ais-eae/plan/index-eng.html

Haber, I. (2020, August 18). *Invasive 2 Species Take Place Among Mediterranean Fish*. İklim Haber. https://ww.iklimhaber.org/istilaci-2-tur-akdeniz-baliklari-arasindaki-yerini-aldi/

IUCN [International Union For Conservation Of Nature]. (n.d.). *Marine Menace, Alien invasive species in the marine environment*. IUCN. https://www.cbd.int/invasive/doc/marine-menace-iucn-en.pdf

Katsanevakis, S., Coll, M., Piroddi, C., Steenbeek, J., Ben Rais Lasram, F., Zenetos, A., & Cardoso, A. C. (2014). Invading the Mediterranean Sea: Biodiversity patterns shaped by human activities. *Frontiers in Marine Science*, *1*, 32. doi:10.3389/fmars.2014.00032

Katsanevakis, S., Tempera, F., & Teixeira, H. (2016). Mapping the impact of alien species on marine ecosystems: The Mediterranean Sea case study. *Diversity & Distributions*, *22*(6), 694–707. doi:10.1111/ddi.12429

Kourantidou, M., Cuthbert, R., Haubrock, P., Novoa, A., Taylor, N., Leroy, B., Capinha, C., Renault, D., Angulo, E., Diagne, C., & Courchamp, F. (2021). Economic costs of invasive alien species in the Mediterranean basin. *NeoBiota*, *67*, 427–458. doi:10.3897/neobiota.67.58926

Mannino, A. M., & Balistreri, P. (2021). Invasive alien species in Mediterranean Marine Protected Areas: The Egadi Islands (Italy) case study. *Biodiversity (Nepean)*, *22*(1-2), 13–23. doi:10.1080/14888386.2021.1875263

Marchini, A., Galil, B. S., & Occhipinti-Ambrogi, A. (2015). Recommendations on standardizing lists of marine alien species: Lessons from the Mediterranean Sea. *Marine Pollution Bulletin*, *101*(1), 267–273. doi:10.1016/j.marpolbul.2015.09.054 PMID:26471066

Marignani, M., Lussu, M., Murru, V., Bacaro, G., & Cogoni, A. (2020). Effect of Invasive Alien Species on the Co-Occurrence Patterns of Bryophytes and Vascular Plant Species—The Case of a Mediterranean Disturbed Sandy Coast. *Diversity (Basel)*, *12*(4), 160. doi:10.3390/d12040160

NOAA Fisheries [National Oceanic and Atmospheric Administration]. (n.d.). *Invasive and Exotic Marine Species*. NOAA. https://www.fisheries.noaa.gov/insight/invasive-and-exotic-m arine-species

Occhipinti-Ambrogi, A. (2007). Global change and marine communities: Alien species and climate change. *Marine Pollution Bulletin*, *55*(7-9), 342–352. doi:10.1016/j.marpolbul.2006.11.014 PMID:17239404

Occhipinti-Ambrogi, A., Marchini, A., Cantone, G., Castelli, A., Chimenz, C., Cormaci, M., Froglia, C., Furnari, G., Gambi, M. C., Giaccone, G., Giangrande, A., Gravili, C., Mastrototaro, F., Mazziotti, C., Orsi-Relini, L., & Piraino, S. (2011). Alien species along the Italian coasts: An overview. *Biological Invasions*, *13*(1), 215–237. doi:10.100710530-010-9803-y

Özacar, İ. U. (2019, September 26). *Fish stocks are declining in the seas*. TRT Haber. https://www.trthaber.com/haber/turkiye/denizlerde-balik-stok u-azaliyor-432836.html

Özacar, İ. U. (2020, September 2). *Climate change is also increasing invasive species in the seas*. TRT Haber. https://www.trthaber.com/haber/turkiye/iklim-degisikligi-den izlerdeki-istilaci-turleri-de-artiriyor-513312.html

Sarı, İ. (2021, September 13). *Invasive species threaten Marmara and Black Sea as well! What awaits us if native fish species decrease?* Hurriyet. https://www.hurriyet.com.tr/gundem/istilaci-turler-marmara-v e-karadenizi-de-tehdit-ediyor-yerli-balik-turleri-azalirsa-b izi-neler-bekliyor-41885358

Servello, G., Andaloro, F., Azzurro, E., Castriota, L., Catra, M., Chiarore, A., Crocetta, F., D'Alessandro, M., Denitto, F., Froglia, C., Gravili, C., Langer, M. R., Lo Brutto, S., Mastrototaro, F., Petrocelli, A., Pipitone, C., Piraino, S., Relini, G., Serio, D., & Zenetos, A. (2019). Marine alien species in Italy: A contribution to the implementation of descriptor D2 of the marine strategy framework directive. *Mediterranean Marine Science*, *20*(1), 1–48. doi:10.12681/mms.18711

The Fish Site. (2021, April 28). *Turkey breaks $1 billion aquaculture export milestone*. The Fish Site. https://thefishsite.com/articles/turkey-breaks-1-billion-aquaculture-export-milestone

TURKSTAT. (2021a). *Fishery statistics*. TurkStat. https://data.tuik.gov.tr/Kategori/GetKategori?p=tarim-111&dil=1

TURKSTAT. (2021b). *Quantity of caught sea fish*. TurkStat. https://data.tuik.gov.tr/Kategori/GetKategori?p=tarim-111&dil=1

TURKSTAT. (2021c). *Quantity of caught other sea fish (crustaceas, molluscs)*. TurkStat. https://data.tuik.gov.tr/Kategori/GetKategori?p=tarim-111&dil=1

TURKSTAT. (2021d). *Landing of sea fish for intended use, 2020*. TurkStat. https://data.tuik.gov.tr/Kategori/GetKategori?p=tarim-111&dil=1

TURKSTAT. (2021e). *Landing of other sea products for intended use, 2020*. TurkStat. https://data.tuik.gov.tr/Kategori/GetKategori?p=tarim-111&dil=1

WWF. (2018). *Puffer fish*. World Wildlife Foundation. https://www.wwf.org.tr/basin_bultenleri/guncel_haberler/?7420/balonbaliklari

Yildirim, S., Demirtas, I., & Yildirim, D. C. (2022). A Review of Alternative Economic Approaches to Achieve Sustainable Development: The Rising Digitalization and Degrowth Post COVID-19. In R. Castanho (Ed.), *Handbook of Research on Sustainable Development Goals, Climate Change, and Digitalization* (pp. 288–307). IGI Global. doi:10.4018/978-1-7998-8482-8.ch018

Yildirim, S., & Kaplan, M. (2020). Seafood Security and Sustainability Through Sustainable Development: A Review of Turkish Seafood Market. In A. Obayelu & O. Obayelu (Eds.), *Developing Sustainable Food Systems, Policies, and Securities* (pp. 16–36). IGI Global., doi:10.4018/978-1-7998-2599-9.ch002

Yildirim, S., & Kaplan, M. (2022). The Threat of Invasive Alien Marine Species to the Blue Economy: The Mediterranean Case. In Raimi, L., & Kah, J. M. (Eds.). Implications for Entrepreneurship and Enterprise Development in the Blue Economy. IGI Global. doi:10.4018/978-1-6684-3393-5.ch004

Yıldırım, S., & Yıldırım, D. C. (2021). Achieving Seafood Security in the Mediterranean Region: A Case of Turkey. In R. Castanho & J. Martín Gallardo (Eds.), *Management and Conservation of Mediterranean Environments* (pp. 175–195). IGI Global., doi:10.4018/978-1-7998-7391-4.ch011

Yıldırım, S., Yıldırım, D. Ç., & Gedikli, A. (2018). Sustainable Consumption Trends in the World in the Context of Green Economy and Sustainability. In I. Management Association (Ed.), Sustainable Development: Concepts, Methodologies, Tools, and Applications (pp. 1605-1624). IGI Global. doi:10.4018/978-1-5225-3817-2.ch071

Zenetos, A., Gofas, S., Morri, C., Rosso, A., Violanti, D., Raso, J. G., & Verlaque, M. (2012). Alien species in the Mediterranean Sea by 2012. A contribution to the application of European Union's Marine Strategy Framework Directive (MSFD). Part 2. Introduction trends and pathways. *Mediterranean Marine Science*, *13*(2), 328–352. doi:10.12681/mms.327

ADDITIONAL READING

Esen, Ö., Yıldırım, D. Ç., & Yıldırım, S. (2020). Threshold effects of economic growth on water stress in the Eurozone. *Environmental Science and Pollution Research International*, *27*(25), 31427–31438. doi:10.100711356-020-09383-y PMID:32488700

Yildirim, S., & Kaplan, M. (2020). Seafood Security and Sustainability Through Sustainable Development: A Review of Turkish Seafood Market. In A. Obayelu & O. Obayelu (Eds.), *Developing Sustainable Food Systems, Policies, and Securities* (pp. 16–36). IGI Global., doi:10.4018/978-1-7998-2599-9.ch002

Yıldırım, S. & Kaplan, M. (2020). The Profile of Turkey's Sea and Seafood in the context of Sustainable Development. *Gaziantep University Journal of Social Sciences, 19*(3), 853-866.

Yildirim, S., & Kaplan, M. (2022). The Threat of Invasive Alien Marine Species to the Blue Economy: The Mediterranean Case. In Raimi, L., & Kah, J. M. (Eds.). Implications for Entrepreneurship and Enterprise Development in the Blue Economy. IGI Global.

Yıldırım, S., & Yıldırım, D. C. (2021). Achieving Seafood Security in the Mediterranean Region: A Case of Turkey. In R. Castanho & J. Martín Gallardo (Eds.), *Management and Conservation of Mediterranean Environments* (pp. 175–195). IGI Global., doi:10.4018/978-1-7998-7391-4.ch011

Yıldırım, S., Yıldırım, D. Ç., & Gedikli, A. (2018). Sustainable Consumption Trends in the World in the Context of Green Economy and Sustainability. In I. Management Association (Ed.), Sustainable Development: Concepts, Methodologies, Tools, and Applications, (pp. 1605-1624). IGI Global. doi:10.4018/978-1-5225-3817-2.ch071

KEY TERMS AND DEFINITIONS

FAO: Food and Agriculture Organization of the United Nations **Invasive alien species:** These species spreads outside and then threatens local/natural biological diversity. **Seafood Market:** A type of food market including seafood products.

Seafood Products: Seafood products include fish and others that provided from marines and oceans.

Sustainable seafood market: When a seafood market provides sufficient quantity of several seafood products for the demand, it can be called as sustainable seafood market.

TURKSTAT: Turkish Statistical Institute

Section 4

Hydrogeological and Climatic Aspects of Wetlands: Past and Current Changes

Chapter 6
Water Crisis in Tunisia During the Anthropocene and Great Acceleration Wetlands as Key Sites of Hydrogeological Modifications

Elhoucine Essefi
Faculty of Sciences of Sfax, University of Sfax, Tunisia

Soumaya Hajji
https://orcid.org/0000-0002-6371-7223
LR3E, ENIS, University of Sfax, Tunisia

ABSTRACT

Due to overexploitation, anthropogenic pollution, and climatic change during the Anthropocene and Great Acceleration, Tunisia has already witnessed a severe water crisis. Since they are the natural outlets of their hydrological and hydrogeological basins, wetlands are candidates to obviously display the radical change in water cycles related to anthropogenic activities. This chapter provides efficient methods to investigate water cycle modifications through the study of the water budgets within wetlands (Sidi El Hani Discharge Playa, Eastern Tunisia was taken as example). As concrete manifestation of this change, the dryness of in water springs or the decrease of their flows are concrete indicators of a real water crisis in Tunisia. In addition, salt consumption leads to a decreasing salinization.

DOI: 10.4018/978-1-7998-9289-2.ch006

INTRODUCTION

Due to increasing population, pollution and climatic changes during the Anthropocene and Great Acceleration (Hajji et al., 2018, 2020), Tunisia is one of the African countries threatened by sacristy of water resources. In the coming decades, Tunisia is candidate to face a real water crisis (Hajji et al., 2021). As a matter of fact, Anthropocene conditions would have serious repercussions deepening difficulties. On the one hand, climate change marked by decreasing rains along the country influenced the available surface water stoked within dams and artificial lakes (Wolanski and Hopper, 2022). In addition, the recharge of shallow aquifer decreased (Farid et al., 2022). On the other hand, due to an increasing population, which results in an overexploitation and pollution of surface and subsurface water resources, is related to the setting of the Anthropocene. During the Anthropocene and Great Acceleration, climate change and the subsequent environmental impact havebecome of particular international interest; it has an impact on water recourses collected in wetlands (e.g., Brinkman, 2021; TekadeRupali, 2021). As for the Tunisian context, the connectivity between groundwater and saline environments in the Tunisian Sahel area makes these bodies little more than a surface expression of the water table (Essefi, 2009; Essefi et al., 2013, 2014a). Thus, the impact of climatic variability during the Anthropocene and Great Acceleration in saline systems should be viewed within a framework of a combined groundwater-wetland system (Carol et al., 2022). This chapter is meant to gain an insight into the sensitivity of these systems to climatic change during the Anthropocene and Great Acceleration. Nevertheless, it will not predict discharge playa or groundwater environment global changes, nor will it predict how changes in climate will affect its entire hydrological cycle. Rather, it will provide an insight into how playas, which are strongly influenced by a regional salty groundwater flow regime, such as Sidi El Hani discharge playa, would respond to the increasing dryness taking place during the Anthropocene and Great Acceleration. That is to say, a reinforcing of the idea previously discussed in other works (Essefi et al., 2014b) of a strong deep influence of somewhere domes of salt and/or brinecoming from the Triassic and/or Messinian Salinity Crisis MSC formations located in the subsurface of the Sahel area.According to the proposed model, this dominance of the groundwater contribution may be increased by the installation of dams in the Tunisian center, which increases the convergence of Kairouan aquifer towards Sidi El Hani discharge playa; and hence increases its salinization. Here, we can find out the double contradictory effects of Anthropocene and Great Acceleration changes. On the one hand, arid conditions enhance the scarcity of water resources and salinization within wetlands. On the other hand, dams setting (especially during the Great Acceleration) decreases water collected from its hydrological basin and increases water collected from its hydrogeological basin.

METHODS

Two strategies were followed in the study of the groundwater contribution in Sidi El Hani discharge playa to propose a model relating Sidi El Hani discharge playa and the surrounding aquifers during the Anthropocene and Great Acceleration (e.g., Kairouan aquifer): (1) a hydrological and hydrogeological approach; (2) a geochemical approach. The first was meant to be applied within the discharge playa itself or within its hydrogeological watershed by selecting between the following different hydrological and hydrogeological methods: the field oriented approach, the numerical simulation and the balance technique. The second recognized, on the basis of the first strategy, that Sidi El Hani discharge playa is commanded mainly by its hydrogeological basin rather than its hydrological watershed; and it studied the geochemistry of the converging water. In order to take aside the hydrological contribution, this study was meant to be in some drills outside the discharge playa.

Hydrogeological and Hydrological Methods for Estimating the Groundwater Contribution and Their Efficiency in Sidi El Hani Discharge Playa

Even though they are two different factors, effects of hydrology and hydrogeology are hardly distinguished in the case of a discharge playa, because the limit between the surface and subsurface water is not always quite evident. For instance, the wetland of Sidi El Hani is fed by the hydrogeology through the permanent hydrological network, Oued Chrita. In this study, the focus was first on the selection of the most efficient method to deal with Sidi El Hani case. To estimate groundwater flow into, or from, the wetland, three hydrogeological and hydrological methods may be used. In this study, the selected method should be compatible with the context of this discharge playa and with the available data; accordingly, advantages and drawbacks of each method, then, were discussed before judging the convenient method.

Field-oriented Approach

The field-oriented method is used to interpret potential pathways for groundwater flow into, or from the studied playa (Winter, 1986). In this method, deep or shallow piezometers are installed around the wetland. At the bottom of the wetland, a seepage meters may be also placed. To quantify groundwater flux, values of hydraulic head, hydraulic conductivity and hydraulic gradient obtained from piezometers are applied through the Darcy equation (e.g., Rhinaldo-Lee & Anderson, 1980). Indeed, accurate data on groundwater and solute flux may be obtained by using seepage meters. But these data are relevant only on a local scale. Because, when attempting to extrapolate these data over even small lake-watersheds, estimations are not accurate because of the considerable spatial variability of the subsurface structure and permeability, the temporal and spatial variation of the flow rates and the hydraulic gradient. Extrapolating the point-specific measurements of groundwater discharge obtained from seepage meters to an entire playa, then, will lead to errors, especially if the playa is relatively large; which is the case of Sidi El Hani discharge playa (370 km2). In the present study, it was partially used in order to qualitatively investigate the groundwater contribution and to prove that the convergence of the surrounding aquifers towards Sidi El Hani discharge playa may be commanded by the groundwater system. This use was motivated by the available hydrogeological map (Figure 1) discussing the surrounding aquifers of Sidi El Hani discharge playa.

Numerical Simulation

Numerical modeling of groundwater-flow systems in the vicinity of a playa is used for understanding the controlling parameters and processes affecting groundwater flow near a Salt Lake. However, most existing numerical modeling studies are mainly limited to hypothetical scenarios (e.g., Winter & Pfannkuch, 1984); because, required detailed field data to accurately characterize the spatial and temporal variations in playa-basin are not always available. Models of lake-watersheds have been used to provide a quick assessment of groundwater flux under a variety of hydrological conditions (e.g., Munter & Anderson, 1981); but watersheds considered in this method are very small (few km2). It is worth to be noted that in the present-day Sidi El Hani watershed is approximately 3300 km2. Thus, it was better to take away this method.

Hydrologic Balance of the Playa-watershed

Hydrologic balance is also called the budget technique. It is the most common method to assess the entering and leaving salt and water in the system. It has been widely used for simulating, and predicting, runoff and stream flow since the late 1950s (e.g., Cherkauer & Nader, 1989). Balance techniques are based on accounting of the hydrological and hydrogeological contributions within the saline environment itself and ignore the spatial variability of the physical components of its watershed.

Hydrologic balance studies are based on quantifying the volume of water or the variation of the salt budget in the saline environment due to precipitation, evaporation, surface runoff, lake discharge, and groundwater inflow or outflow. Contributions to the volume of water or salt within the saline environment from the hydrological and hydrogeological components, other than the groundwater components, can easily be measured; or, in the case of evaporation, can be calculated with commonly used formulae (e.g., Gray et al., 1970); but in Sidi El Hani discharge playa the actual evaporation itself is influenced by the salt budget (fluid density) (Essefi et al., 2013); the actual evaporation from the discharge playa is quite different from the potential evaporation given from the meteorological station. Thus, evaporation is to be corrected by the fluid density. Advantages of the balance approach compared to the previous methods liein its estimation of the net groundwater contribution without using detailed field data. But before applying the method of salt budged technique in this study about Sidi El Hani discharge playa, the theoretical background describing the modeling techniques, based on the water balance and salt balance should be presented.

Water Balance Equation

It is used to estimate the water budget within the wetland. As first impression, groundwater inflow may be considered as the feeder of Sidi El Hani wetland. The permanent surface of brine during the summer in a deficit hydrologic system and relatively small drainage basin should be a result of a water coming up. The quantification of this water budget needs an equation that collects all components participating in addition or removal of water. The balance equation is:

$$Qm=Qr+ Qf +Qsw+ Qgd + Qgs- Qe \quad (1)$$

(Yechieli & Wood, 2002)

Where Qm(L3/T) is the average outflow from the saline environment to the ocean (L3/T) (L3: international unit of volume; M: international unit of time), Qsw(L3/T) is the average recharge due to marine flooding (L3/T). Since Sidi El Hani is an inland sebkha, components related to sea (Qm=0 andQsw=0) in equation (1) are omitted. Qr(L3/T) is the average rainfall (L3/T). Qf is the average overland flooding onto the playa (L3/T). Qgs is average shallow groundwater inflow to the playa (L3/T); Qgd is the average deep groundwater inflow to the playa (L3/T). These very two last components referring to groundwater contribution (Qgs and Qgd) are usually deduced from this equation (1). They would be separately estimated. But because of a complicated hydrogeological situation, it would be better to deal with (Qg = Qgd + Qgs) (L3/T) as the average total groundwater inflow. The Figure 1 shows that water inflows converge toward the discharge playa from different aquifer systems; the plain of Kairouan drains northward to the outlet of Kelbia and southward to Sidi El Hani in the North (Hollis & Kallel, 1985); and the groundwater converges toward the discharge playa from Souassi and Zarmdine. Qe(L3/T) is the net evapotranspiration(L3/T); actually, not only in the case of Sidi El Hani discharge playa but also for any saline environment, the actual evaporation is different from the potential evaporation obtained from pitch condition. As a matter of fact, evaporation with a distilled water could by no means be equivalent to an evaporation taking place in a water with a salinity exceeding 350 g/L. Hence, it would be safer to speak in this equation about Qea as the actual evapotranspiration. So, the equation (1) becomes:

$$Qea = Qr + Qf + Qg \quad (2)$$

(Yechieli & Wood, 2002; modified)

Solutes Balance Equation

Generally, solutes in playas are derived from rain, flooding by seawater, overland flooding, shallow ground-water discharge, and deep ground-water discharge (hence the name of discharge playas). The equation may be written:

$$QmCm = Qea + Qr\,Cr + QswCsw + QfCf + QgsCgs + QgdCgd - G - SP \pm E \quad (3)$$

(Yechieli & Wood, 2002)

Where Cm is average solute concentration in the outflow to the sea (M/L3) (M: international unit of mass; L3: international unit of volume); Cr is the average solute concentration in the rain (M/L3); Csw is the average solute concentration of the marine flooding (M/L3); Cf is the average solute concentration of overland flooding (M/L3); Cgs is the average solute concentration of the shallow groundwater inflow (M/L3); Cgd is the average regional solute concentration of the deep ground-water inflow (M/L3); drilling data (see the hydrochemistry of some drills in the vicinity of Sidi El Hani wetland) give Cgs= Cgd. SP is the mineral precipitation rate and water–rock interaction processes (M/T), G is the rate of conversion of solute to gas and radioactive decay (M/T), and E is Aeolian and other erosion processes (M/T); all other parameters (Q terms) were defined. Nevertheless, in the very saline brine such as of Sidi El Hani discharge playa, the unit of mass/mass would be more appropriate. Taking into account the naught values mentioned in the previous paragraph, the equation (3) could be replaced by the following one:

$$QfCf = Qea + Qr\,Cr + Qg\,Cg - G - SP \pm E \quad (4)$$

(Yechieli & Wood, 2002; modified)

In Sidi El Hani case, another mechanism of acquiring salt is noticed: dissolution (D); since the ever present salt crust tends to be thinner during the wet seasons (winter); that is to say, salt reserve in the wetlandincrease the fluid density. To be sure, it may be encompassed in the term E (other processes) of Yechieli & Wood equation (2002). But in Sidi El Hani case, this component is worth to be mentioned as a standalone component: D; hence, we can write:

$$QfCf = Qea + Qr\ Cr + Qg\ Cg - G - SP \pm E + D \qquad (5)$$

(Yechieli & Wood, 2002; modified)

To give verbal voice to this equation, the salt budget is commanded by the groundwater contribution, the precipitation, the dissolution, the evaporation and other processes. Taking aside some components (G, P, E and D) at this stage, this equation will be the guide line to compile the cause-effect (Ichikawa) diagram of the fluid density and of the actual evaporation in Sidi El Hani discharge playa and to deduce the relation between them.

Application of the Appropriate Methods to Estimate Groundwater Contribution in Sidi El Hani Discharge Playa

As it is previously commented, two hydrological and hydrogeological methods seem convenient to the case of Sidi El Hani saline environmentField oriented-approach. In the case of Sidi El Hani saline environment, the field oriented-approach is applied to decide the way of water flow in the subsurface. The hydrogeological map of the surrounding aquifers (Figure 1) is completed to prove that water converges towards the discharge playa. The water table within the discharge playa is measured during field missions by coring the discharge playa sediment.

Geochemical Detection of the Groundwater Contribution in Some Drills in the Vicinities of Sidi El Hani Discharge Playa

Analyses of water of ten drills (Figure 1) recuperated from data of the Ministry of Agriculture were taken to endure a geochemical study. First, Piper diagram studies the geochemical facies of samples. Second, the use of Schoeller-Berkalof diagram was meant to prove relatedness between drills in order to understand if Sidi El Hani discharge playa is fed by homogenous or different aquifers. Third, the index of mineralization of the groundwater was calculated to explain the origin of the huge quantities of halite in the discharge playa.

RESULTS AND DISCUSSION

The groundwater contribution should be proven; then, we can deal with the quantification of its contribution. Thus, the quality of the convergent water may be geochemically studied.

Qualitative Investigation of the Groundwater Contribution by Field-Oriented Approach

The hydrogeological map (Figure 1) shows that the surrounding aquifers converge towards the saline environment; hence, its name according to the hydrogeology is discharge playa. Thus, the permanent layer of brine, during hot seasons, may be explained. In this vein, one can notice the different hydrogeological behavior of the saline environment of Kelbiaplayalake, which represents a transition zone towards the sea. Based on the hydrogeological map, the surrounding aquifersare governing the wetland history. The salt budget and its relation with the different climatic factors might reinforce the idea of an inherited discharge playa from the Triassic intrusions and/or the Messinian Salinity Crisis. Monitoring fault (cold and hot) and artesian springs has been proven relevant to understand combined variability of the internal and external geodynamics (Finley, 2020). On the one hand, the hydro chemical investigation of artesian springs proves vital to monitor the groundwater flow, which is in terns related to the external earth geodynamics (climate variability and anthropogenic activities). On the other hand, the formation and functioning of fault springs are mainly controlled by both external and internal (tectonic alignments; mantle fluids contribution) (Sato et al., 2020). In arid and semi-arid regions in North Africa, groundwater resources are used for agricultural irrigation. As a result, aquifers are increasingly exploited in order to meet rising agricultural demands (Mokadem et al., 2018). Groundwater depletion, soil salinization, and even desertification are all consequences of extensive groundwater extraction (Mamou & Kassah, 2002). During the Anthropocene and Great Acceleration, the overexploitation resulted from a decreasing trend in precipitation, related to a climatic change and an increasing groundwater extraction due to an increasing population have caused critical drop in piezometric levels, water-quality degradation, increased abstraction costs, ecological damage, increased pumping costs, and decreased well yields (Konikow, 2002).

Geochemical Investigation of the Groundwater Contribution in Sidi El Hani Discharge Playa

Piper Diagram

Concerning Piper diagram (Figure 2), the interpretation of the dispersion of different samples within the lozenge of anions and cations, the diagram of anions and the diagram of cations may give an idea of the geochemical facies. On the lozenge, three families of samples are distinguished. First, the majority of samples make up the Chlorinated and/or Sodic Sulphated and/or Potassic family (F1). Second, samples D4 and D9 represent the Chlorinated and/or Sodic Sulphated and/or Magnesian family

Figure 1. Geographical location and hydrogeological context of the MCSH system: Groundwater upwelling through springs and spring mounds

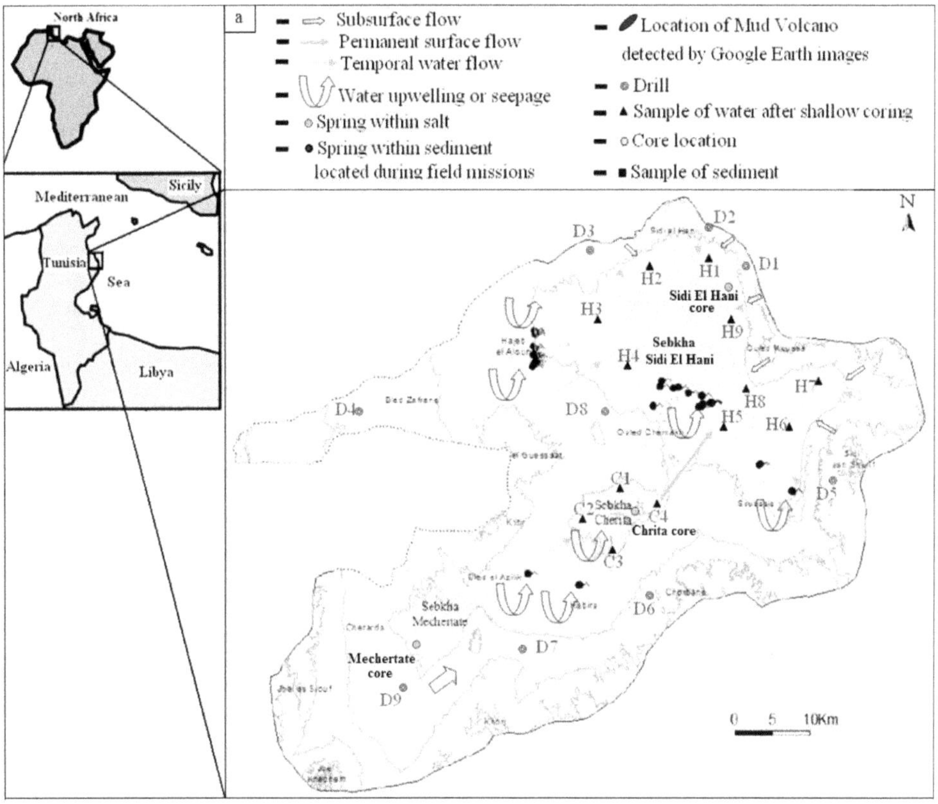

(F2). Third, the sample D67 represents the calcic Bicarbonate and/or Magnesian family (F3). Generally, all drills have the same facies (F1), they are made up of water charged with Na+ and Cl-. This geochemical facies suggest that this aquifer is more than likely leaching domes of halite (Triassic and/or Messinian) or is made up of salty paleo-water probably inherited from the Messinian Salinity Crisis. These findings reinforce the economic importance of Sidi El Hani wetland discussed by Tagorti (1995). Nevertheless, the geochemistry of F2 may be explained by looking back to the geodynamic context; D4 and D9 are located near the Triassic dome in the subsurface of Zarmdine

Schoeller-Berkalof Diagram

As for the studied samples, they present for the most an evident similarity (Figure 3). Geochemical families already distinguished by Piper diagram are also noticed in this diagram. The F1 is marked by high value of Na+ and Cl-. On the other hand, the F2 is marked by high values of SO4--. The F3 is characterised by high value of CO32--. So, samples represent three different aquifers.

Figure 2. Piper diagram of some drills in the vicinities of Sidi El Hani discharge playa

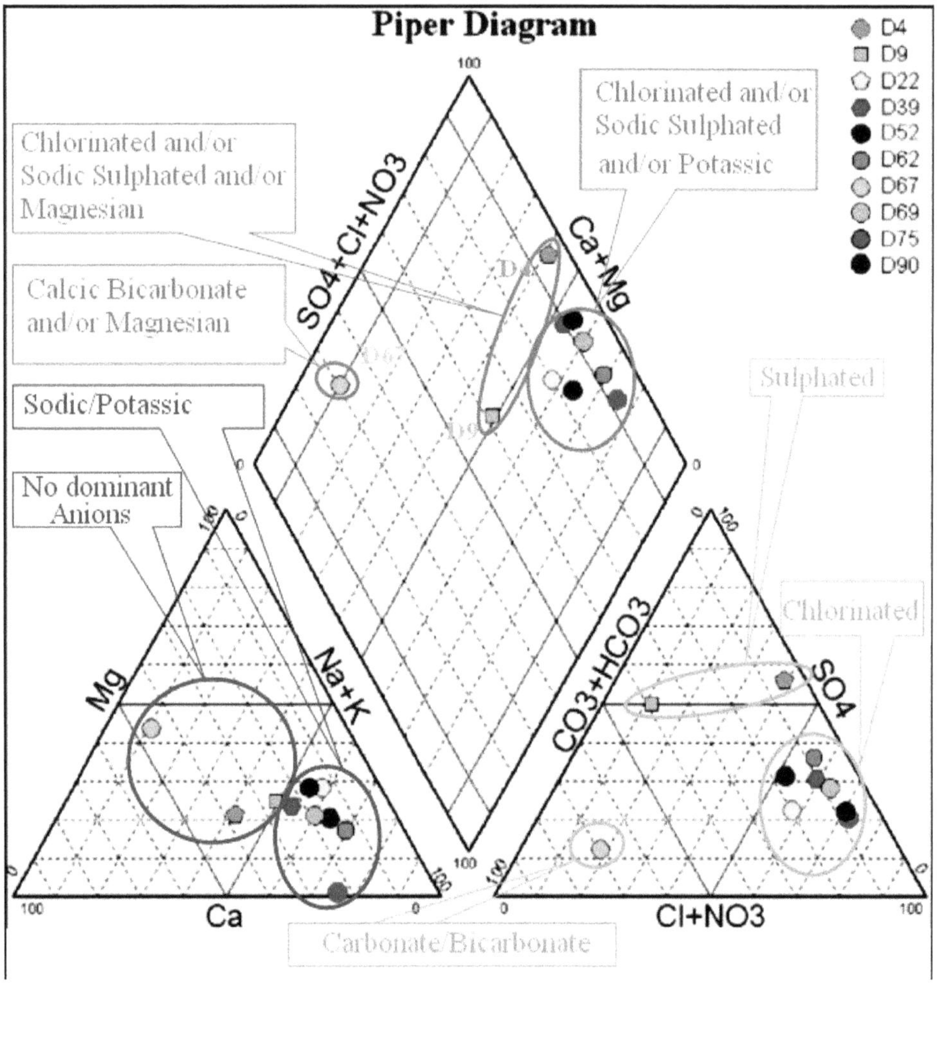

Figure 3. Schoeller-Berkalof diagram of some drills in the vicinities of Sidi El Hani discharge playa

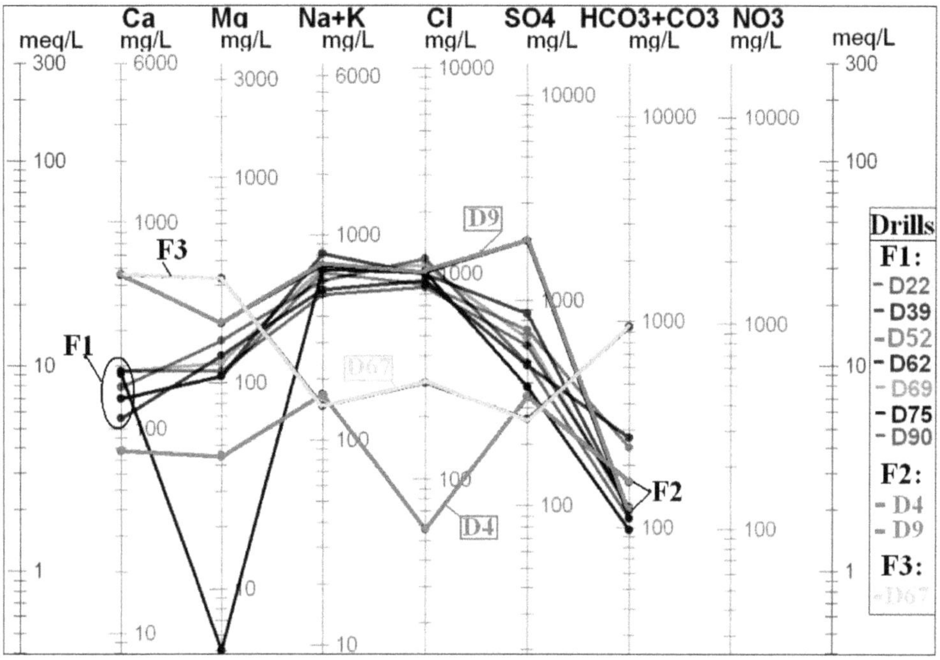

Mechanism of Groundwater Mineralization: Saturation Indices Towards Some Evaporitic Minerals

Index of Saturation Towards Anhydrite and Gypsum

The Figure 3 and Figure 4 show that gypsum and anhydrite are correlated; even though their indices have not the same values; they have the same curves shape. These values are in favour of an undersaturation. Since they are more negative, values of anhydrite are more telling than gypsum ones. Thus, the leaching of domes of gypsum evaporates is more likely to take place. The Triassic intrusions discussed in the geodynamic study are in some extend verified; because, the leaching of gypsum (connoted to be Triassic) is verified. On the other hand, the hypothesis of the Messinian Salinity Crisis MSC tends to be excluded. Because even though some works (e.g., Sghari & Mercier, 2008) discussed the presence of secondary gypsum within Messinian sediments in the Sahel area, the literature (e.g., Hsü et al., 1973) did not relate the Messinian sediment to gypsum.

Figure 4. Index of saturation towards anhydrite

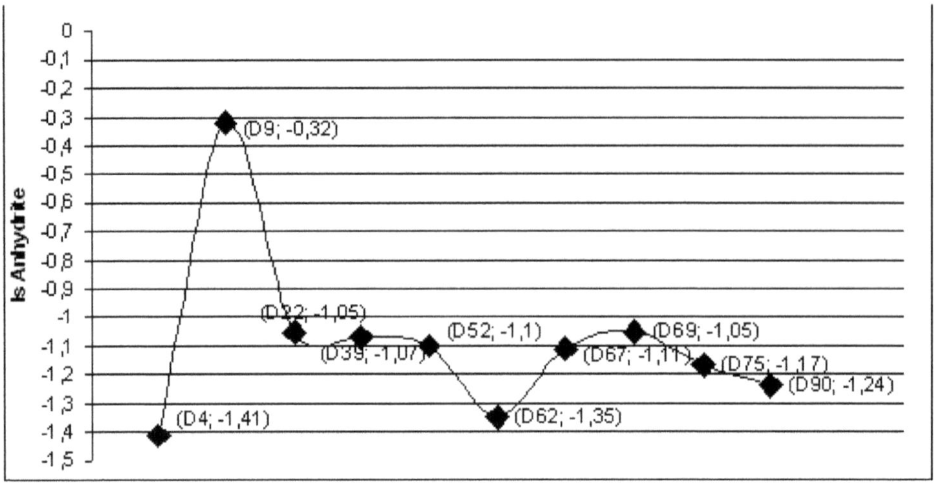

Figure 5. Index of saturation towards gypsum

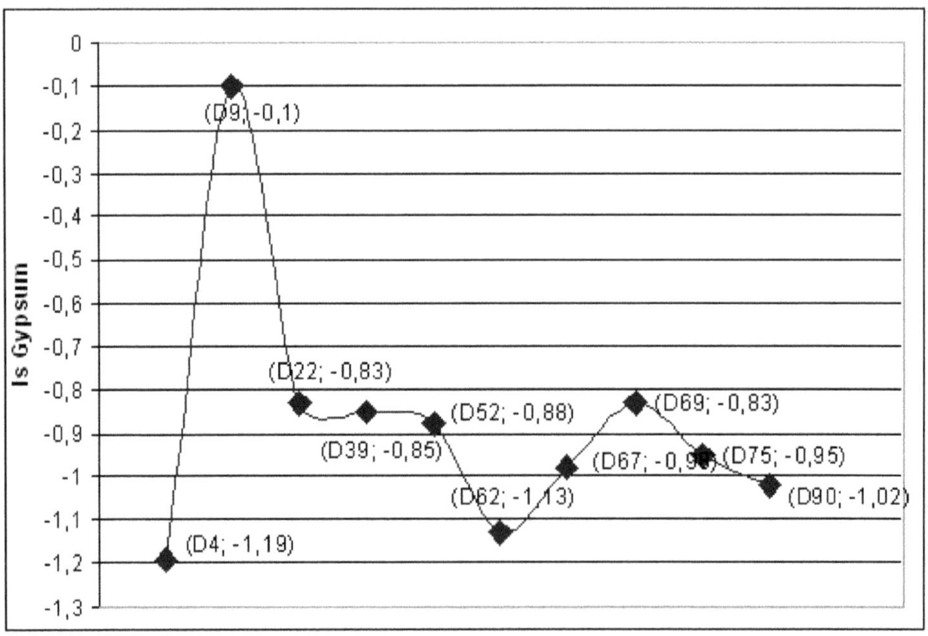

Figure 6. Index of saturation towards halite

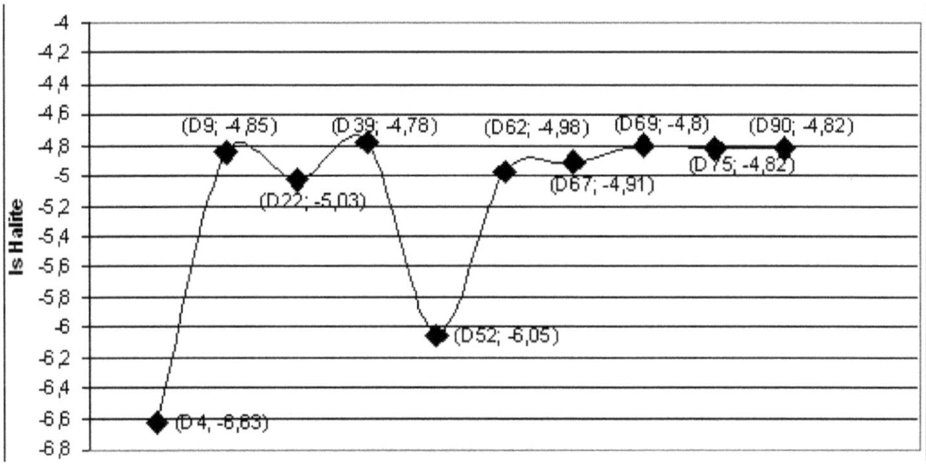

Index of Saturation Towards Halite

As it is quite evident (Figure 5), halite has the lowest (highest eigen values) indices of saturation. Thus, the dissolution of halite is more likely to occur with a coming up of salty water to Sidi El Hani discharge playa. Consequently, the existence of huge quantities of halite in the sebkha and the establishment of the industrial salt pan (SAIIDA), in Sidi El Hani discharge playa, finds all its explanation. Halite may be either Triassic or Messinian. Thus, none of the two possibilities proposed by the geodynamic introduction could be excluded. Only the isotopic study may solve the ambiguity.

A hydrogeological, tectonic and hydrotectonic models were previously proposed (Essefi, 2009; Essefi et al., 2013, 2014b) about the relation of Sidi El Hani saline environment with Kairouan aquifer; the situation with the other surrounding aquifers such as of Souassi and Zarmdine would be alike, namely with the presence of domes of salt in Zarmdine and Sidi Cherif (Bedir, 1989). The Sidi El Hani saline environment is commanded by somewhere deep domes of salt and/or salty water. The upwelling of saline water towards the wetland is expressed by the outcrop of many springs in different sites (Figure 6). It is made clear, hence, for climatologists that whatever the climate changes, never does it affect radically the geochemical behavior and the azoic paleontological content of Sidi El Hani discharge playa.

Figure 7. Spring of water located in the eastern part feeds the discharge playa of Sidi El Hani by an ever coming up of salty water during the summer of 2007

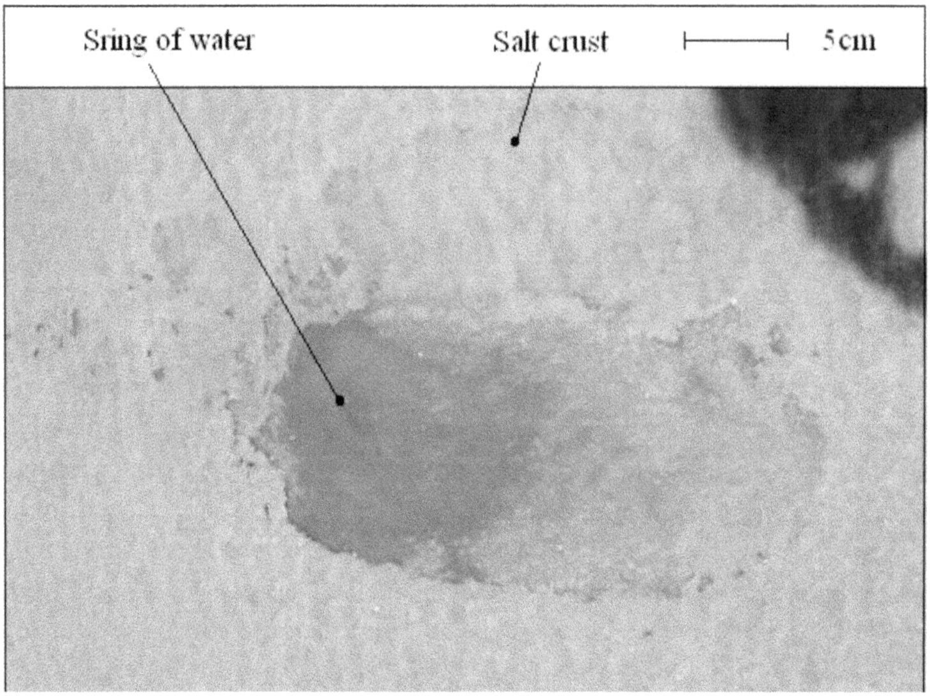

Manifestation of Surface-subsurface Connectivity in Wetlands: Effect of Dryness

As manifestation of aquifers wetland connectivity, Oued Chrita drains permanently even in absence of any rain. This permanent activity of Oued Chrita is strong evidence of water coming up from subsurface. The emanation is expressed by a slow seepage along the the wetland surface. Seepage is enhanced by the permeable sandy sediment.

In the western side of the Sidi El Hani discharge playa, we identified some spring mounds (Figure 8). Their lengths vary from 3 m to 40 m. Their heights may reach a meter. Sometimes, these spring mounds are occupied by central springs including water surges (Figures 8b and 8c). Thus, we can deduce that inactive spring mounds are artesian spring mounds originated by a weak hydraulic pressure. As for the vegetation cover of the spring mounds, plants requiring relatively fresh water occupy the peripheries (Figure 8c and 8d). Closer to the discharge playa (Figure 8d), fresh water-requiring vegetation is substituted for a vegetation tolerating more salinity.

Anthropocene and Great Acceleration are marked by an increasing dryness. Wetlands are key point to record this radical change because they the natural outlets

Figure 8. (a) Outcrop of spring mounds at the surface of the Sidi El Hani wetland and (b), morphology of the spring mound. Sampling (c) and variability of the vegetation with an increasing salinization on an active spring mound from an active spring mound (d) during the summer of 2010

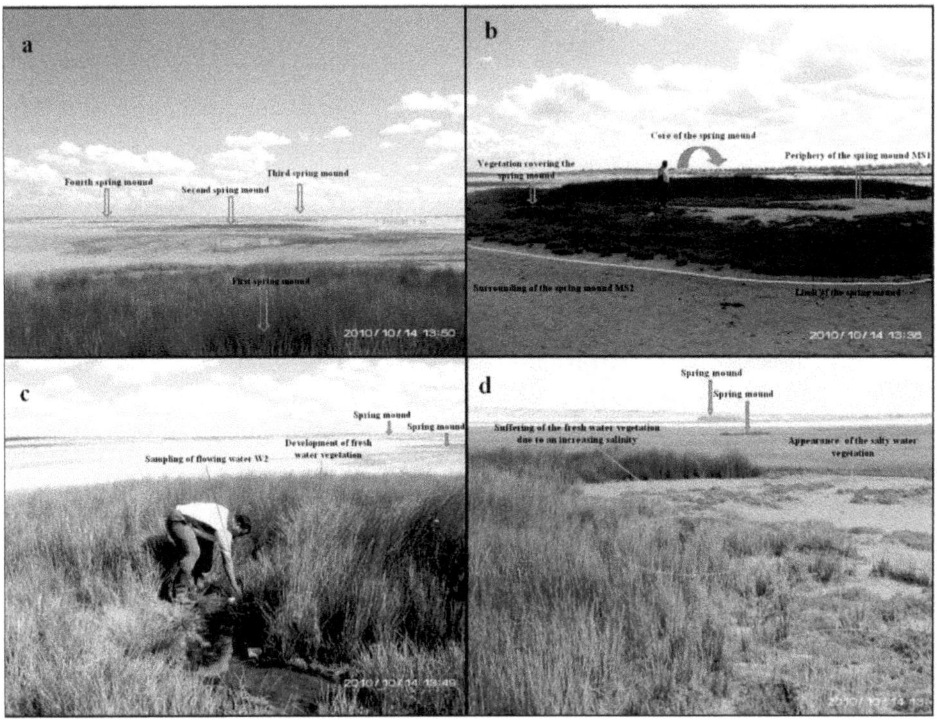

of their hydrological and hydrogeological basin. It is quite evident from the Figures 7, 8 and 9 shows the evolution of the surface of Sidi El Hani wetlands from 2007, 2010 and 2021, respectively. Comparably with Figure 7, in which water obviously outcrops, Figure 9 displays a dried surface of the wetland in the summer of 2021. Further, to reach the water table, we have been obliged to core within the wetland. Here, we can see that wetland are keys sites to display the dryness taking place during the Anthropocene and Great Acceleration.

CONCLUSION

The whole previous analyses show the primary role of the hydrogeological framework and reinforce the idea of an inherited discharge playa from the Messinian Salinity Crisis and/or Triassic evaporitic sedimentation. An ever present high fluid density

Figure 9. Surface of Sidi El Hani wetland in the summer of 2021: an obvious record of climate change during the Anthropocene and Great Acceleration: we have been obliged to core to reach the water table

and (higheigen values) indices of saturation toward halite and sometimes gypsum are the result of convergence of the surrounding aquifers and the leaching of salt formation located in the subsurface of the transporting of brine. The hydrogeological and the geodynamic studies converge toward the same hypothesis. The convergence of the surrounding aquifers resulted in the appearance of spring of water within the discharge playa. These springs prove vital to monitor changes taking place during the Anthropocene and Great Acceleration. These changes include dryness, overexploitation and pollution. Dryness and overexploitation taking place in north Africa resulted in inactivity of water spring or a decrease in their flows. Pollution, especially salinization, is due to human activities related to the setting of the Anthropocene conditions.

REFERENCES

Bedir, M. (1989). *Geodynamics of the sedimentary bassins of Sahel Mahdia (eastern Tunisia), from Aptian to present: sismo-stratigraphy, sismo-tectonics, and sismo-structural* [PhD Thesis]. University of El Manar Tunis.

Brinkman, R. (2021). Impact of climatic change on coastal agriculture. *Climatic Change*, 235–245.

Carol, E., Pilar Alvarez, M. D., Santucci, L., Candanedo, I., & Arcia, M. (2022). Origin and dynamics of surface water-groundwater flows that sustain the Matusagaratí Wetland, Panamá. *Aquatic Sciences*, *84*(1), 1–15. doi:10.100700027-021-00847-y

Cherkauer, D. S., & Nader, D. C. (1989). Distribution of groundwater seepage to large surface water bodies: The effect of hydraulic heterogeneities. *Journal of Hydrology (Amsterdam)*, *109*(1-2), 151–165. doi:10.1016/0022-1694(89)90012-7

Essefi, E. (2009). *Multidisciplinary study of Sidi El Hani Saline Environment: The Geological History and the Climatic Variability*. Academic Press.

Essefi, E., Komatsu, G., Fairén, A. G., Chan, M. A., & Yaich, C. (2014a). Models of formation and activity of spring mounds in the Mechertate-Chrita-Sidi El Hani system, eastern Tunisia: Implications for the habitability of Mars. *Life (Chicago, Ill.)*, *4*(3), 386–432. doi:10.3390/life4030386 PMID:25370379

Essefi, E., Touir, J., Tagorti, M. A., & Yaich, C. (2013). Effect of the groundwater contribution, the climatic change, and the human-induced activities on the hydrological behavior of discharge playas: A case study Sidi El Hani discharge playa, Tunisian Sahel. *Arabian Journal of Geosciences*, *6*(10), 3997–4009. doi:10.100712517-012-0659-6

Essefi, E., Touir, J., Tagorti, M. A., & Yaich, C. (2014b). Geodynamic framework of saline systems in eastern Tunisia: Saline depressions inherited from the Triassic intrusions and/or the Messinian Salinity Crisis. *International Scholarly Research Notices*.

Farid, H. U., Khan, Z. M., Shakoor, A., Mubeen, M., Ayub, H. U., Kanwar, R. M. A., & Bilal, M. (2022). Water Resources in Relation to Climate Change. In *Building Climate Resilience in Agriculture* (pp. 145–166). Springer. doi:10.1007/978-3-030-79408-8_10

Finley, T. (2020). *Fault-hosted geothermal systems in southeastern British Columbia*. Academic Press.

Gray, D. M., Mckay, G. A., & Wigam, J. M. (1970). Energy, evaporation, and evapotranspiration. In D. M. Gray (Ed.), *Handbook on the Principles of Hydrology, Sect. III* (pp. 3.1–3.66). Water Information Center.

Hajji, S., Ayed, B., Riahi, I., Allouche, N., Boughariou, E., & Bouri, S. (2018). Assessment and mapping groundwater quality using hybrid PCA-WQI model: Case of the Middle Miocene aquifer of Hajeb Layoun-Jelma basin (Central Tunisia). *Arabian Journal of Geosciences, 11*(20), 1–21. doi:10.100712517-018-3924-5

Hajji, S., Karoui, S., Nasri, G., Allouche, N., & Bouri, S. (2021). EFA-CFA integrated approach for groundwater resources sustainability in agricultural areas under data scarcity challenge: Case study of the Souassi aquifer, Central-eastern Tunisia. *Environment, Development and Sustainability, 23*(8), 1–20. doi:10.100710668-020-01155-5

Hajji, S., Nasri, G., Boughariou, E., Bahloul, M., Allouche, N., & Bouri, S. (2020). Towards understanding groundwater quality using hydrochemical and statistical approaches: Case of shallow aquifer of Mahdia–KsourEssaf (Sahel of Tunisia). *Environmental Science and Pollution Research International, 27*(5), 5251–5265. doi:10.100711356-019-06982-2 PMID:31848945

Hollis, G.E. & Kallel, M.R. (1985). *Modelling natural and human induced hydrological changes on Sebkhet El Kelbia*. Academic Press.

Hsü, K. J. KJ, H., MB, C., & WBF, R. (1973). The origin of the Mediterranean evaporites. Academic Press.

Konikow, L. F. (2002, October). Ground-water depletion and overexploitation; a global problem. *Geological Society of America Abstracts with Programs, 34*, 229.

Mamou, A., & Kassah, A. (2002). Eau et développement dans le sud tunisien [Water and development in southern Tunisia]. Cahier du CERES, Série Géographique no. 23.

Mokadem, N., Redhaounia, B., Besser, H., Ayadi, Y., Khelifi, F., Hamad, A., Hamed, Y., & Bouri, S. (2018). Impact of climate change on groundwater and the extinction of ancient "Foggara" and springs systems in arid lands in North Africa: A case study in Gafsa basin (Central of Tunisia). *Euro-Mediterranean Journal for Environmental Integration, 3*(1), 1–14. doi:10.100741207-018-0070-0

Munter, J. A., & Anderson, M. P. (1981). The use of ground-water flow models for estimating lake seepage rates. *Ground Water, 19*(6), 608–616. doi:10.1111/j.1745-6584.1981.tb03515.x

Rhinaldo-Lee, M. B., & Anderson, M. P. (1980). High water levels in ground-water dominated lakes- a case study from north-western Wisconsin. *Ground Water*, *8*, 334–339.

Sato, T., Takahashi, H. A., Kawabata, K., Takahashi, M., Inamura, A., & Handa, H. (2020, January 1). Changes in the nitrate concentration of spring water after the 2016 Kumamoto earthquake. *Journal of Hydrology (Amsterdam)*, *580*, 124310. doi:10.1016/j.jhydrol.2019.124310

Sghari, A., & Mercier, J. L. (2008). Le Messinien de la formation "Segui" du Sahel tunisien: un équivalent des "Gilbert deltas" [The Messinian of the "Segui" formation of the Tunisian Sahel: an equivalent of the "Gilbert deltas"]. Séance spécialisée de la Société Géologique de France, 70.

Tagorti, M. A. (1995). Rapport d'étude bibliographique sur le secteur de la sebkha de Sidi El Hani [Bibliographic study report on the sebkha sector of Sidi El Hani]. Academic Press.

TekadeRupali, P. (2021). The Global Climate Changes and Its Impact on Aquatic Environment. International Journal for Innovative Research in Multidisciplinary Field, 282.

Winter, T. C., & Pfannkuch, H. O. (1984). Effect of anisotropy and groundwater system geometry on seepage through lakebeds: 12. Numerical simulation analysis. Journal of Hydrology (Amsterdam), 75(1-4), 239–253. doi:10.1016/0022-1694(84)90052-0 doi:10.1016/0022-1694(84)90052-0

Winter, T. U. (1986). Effect of ground-water recharge on the configuration of the water table beneath sand dunes and on seepage in lakes in the Sandhills of Nebraska. U.S.A. J. Hydrol., 86(3-4), 221–237. doi:10.1016/0022-1694(86)90166-6 doi:10.1016/0022-1694(86)90166-6

Wolanski, E., & Hopper, C. (2022). Dams and climate change accelerate channel avulsion and coastal erosion and threaten Ramsar-listed wetlands in the largest Great Barrier Reef watershed. Ecohydrology & Hydrobiology, 22(2), 197–212. doi:10.1016/j.ecohyd.2022.01.001 doi:10.1016/j.ecohyd.2022.01.001

Yechieli, Y., & Wood, W. W. (2002). Hydrogeologic processes in saline systems: Playas, sabkhas, and saline lakes. Earth-Science Reviews, 58(3-4), 343–365. doi:10.1016/S0012-8252(02)00067-3 doi:10.1016/S0012-8252(02)00067-3

KEY TERMS AND DEFINITIONS

Anthropocene: A new geological epoch proposed by the scientific community for the period were significant.

Climate Change: A climate transformation of global dimension in which there are changes in the ordinary climate of the planet.

Global Warming: The phenomenon of increasing average air temperatures near the surface of Earth over the past one to two centuries.

Great Acceleration: Records the dramatic, continuous, and roughly change in the human activity.

Spring: A natural discharge point of subterranean water at the surface of the ground or directly into the bed of a stream, lake, or sea.

Water Table: The upper level of the subsurface in which the soil or rocks are permanently saturated with water.

Wetlands: Humid zones in which water is permanently or temporally present either by covering soil or incorporating within the near surface of the soil. human impact on Earth's geology and ecosystems has become obvious.

Chapter 7
Geochemistry and Sedimentology Study of a Core From Chott Geuttar:
Central Tunisia Climatic and Environmental Significance

Sana Bedoui
Higher Institute of Water Sciences and Techniques of Gabes, University of Gabes, Tunisia

Elhoucine Essefi
University of Gabes, Tunisia

Younes Hamed
University of Gafsa, Tunisia

ABSTRACT

Endoreic saline systems in Tunisia such as that of Chott El Guettar recorded the climatic changes as well as the setting of millennial cycles. This study deals with high-resolution proxies (colors variability, geochemistry, magnetic susceptibility, and grain size distribution) to find the different cycles. Cycles should be attributed either to internal or external forcing. The output of this study demonstrates a sun-driven Holocene cyclostratigraphy of 1000yr that was previously discussed in the literature. Since color studies were confirmed by direct analyses, this method was used to carry a higher resolution. The solar periodicity clearly identified with a continuous cycle of around 2500 years and with two others of around 1000 and 1660 years only during the Early Holocene. This work confirmed the cycles found by colors study through the real analyses such as geochemical and magnetic measurement.

DOI: 10.4018/978-1-7998-9289-2.ch007

INTRODUCTION

Water and air circulations around the planet change as the climate changes. Nonetheless, some regions immediately respond; others should wait a lapse of time to modify the record of their depositional environments. North Africa is climatically susceptible, and it is the first region in the Mediterranean basin to be affected by climate change. For academic and practical objectives, saline systems have been an important topic of study. Economically, they provide valuable evaporate minerals and brines (Tagorti et al., 2013; Ben Smida, 2016). Sedimentolgical, they give a lot of information about the sedimentary history and the depositional environment (Essefi et al., 2013). To climatologists, the sedimentary records of these systems could be a subject of paleoclimatic (Essefi et al., 2013; 2014; Mefteh et al., 2015) and cyclostratigraphic (Essefi et al., 2014) research.

Sebkhas or chott are considered wetlands. Sebkha is an Arabic word that refers to the salt lakes that are subject to periodic inundations. It is a depression occupied by a generally salty lake and where evaporites are deposited. Throughout human history, wetlands have been considered a valuable natural resource (Scholz and Lee, 2005). 'Wetlands are areas of marsh, fen, peatland, or water, whether natural or artificial, permanent or temporary, with water that is static or flowing, fresh, brackish, or salt, including areas of marine water the depth of which at low tide does not exceed six meters,' according to the Ramsar Convention, which brought wetlands to the attention of the international community (Goodwin, 2017). Wetlands are more prevalent in areas where precipitation exceeds water loss through evapotranspiration and surface runoff (Scholz and Lee, 2005). Moreover, 'Wetlands are a halfway world between terrestrial and aquatic ecosystems and exhibit some of the characteristics of both,' says another, more succinct definition (Smith, 1980). In fact, chott El Geuttar is considered wetlands. It is endorheic system that never have a link to the sea due to the uplifted reliefs avoiding any contact with marine environments (Essefi et al., 2014).

Numerous saline systems have been established in Tunisia. Some of them are endorheic (Essefi, 2009; 2013; Bedoui, 2016; Bedoui et al., 2022); others are exoreic (Lakhdar, 2009). In this part, the studied systems will be introduced in terms of geographic, geologic, and climatic settings. These saline systems are candidates to record the climatic variability. Moreover, the studied saline systems will be introduced in their Mediterranean framework. Then, global phenomena and their effect on North Africa such as the Mediterranean Oscillation (MO), the North Atlantic Oscillation (NAO), and the Atlantic Multidecadal Oscillation (AMO) were introduced to understand the change in El Guettar saline systems with a global line of thinking.

As Complicated areas in which many factors can intervene, saline environments must be studied globally; that is to say that many specialties are involved. The study of saline environments is at the crossroads of many specialties. For this reason, the multidisciplinary study reveals vital in the case of the studied saline system. The study started from the field works to recuperate cores. Then, cores underwent nondestructive analyses such as facies identification and colors study. Once samples were taken from the cores, they underwent several destructive analyses such as geochemical (major elements), geophysical (magnetic susceptibility) and sedimentological (observations, grain size distribution). Statistical and mathematical techniques such as spectral analysis were used to treat the obtained data. The main objective of this study is to find with high resolution the climatic variability in the endorheic saline system.

Study Area and Geologic Setting

Chott El Geuttar is an extensive permanent saline system, located in the semi-arid region of Gafsa, already located on the eastern side of the Mediterranean Sea. It is located between the longitudes 6°30–7°30 east and the latitudes 34°00–34°30 north (Mokadem, 2012; Mokadem et al., 2015, Bedoui et al., 2017). It is a seasonal salt lake on the edge of the Sahara. The alluvial plain of the compartment is occupied by a wetland with an area of 7400 ha (or 8% of the area of the delegation) the center of which spreads Chott El Guettar an area of 3730 ha (Figure 1). This chott occupies the syncline between Jebel Orbata and Berda. Thus, the El Guettar saline system is considered an inland system. It collects water from wadis emanating from Jebel Orbata and Jebel Hachichina. Hydrological, due to the great climatic variability and especially the interannual differences in the rainfall, it can sometimes remain dry for several years. Hydrogeologically, the sebkha is out of reach of groundwater influence because the water table is not next to the surface. However, the eastern part of the sebkha may have signs of groundwater influence, which increases the salinization of the gleysol. This salinization through groundwater involvement proves vital to save its floor from erosion (Essefi et al., 2014). The El Guettar saline system is part of the meridional Atlas of Tunisia. It corresponds to the limit of two paleo-geographic domains which differ in their styles of deformation the Saharan domain and the central Atlas. It is worth noting that the most developed structures are folds with major direction NE-SW, even the existence of some other directions such as E-W, NW-SE is due to abnormal contact (fault of Gafsa) (Hamed et al., 2011).

Indeed, the geology of the El Guettar saline system is characterized by Quaternary clay sediments. At least in the eastern part, salt crust develops on the surface only during the dry season, in the western part; the geomorphologic conditions are closer to those of a sebkha with sandy loam soils. Therefore, the El Guettar saline system is a lower depression, which grows on alluvial, gypsum material with a medium and heavy texture more or less waterlogged.

Figure 1. Geological map of Gafsa (Geological map 1/500 000)

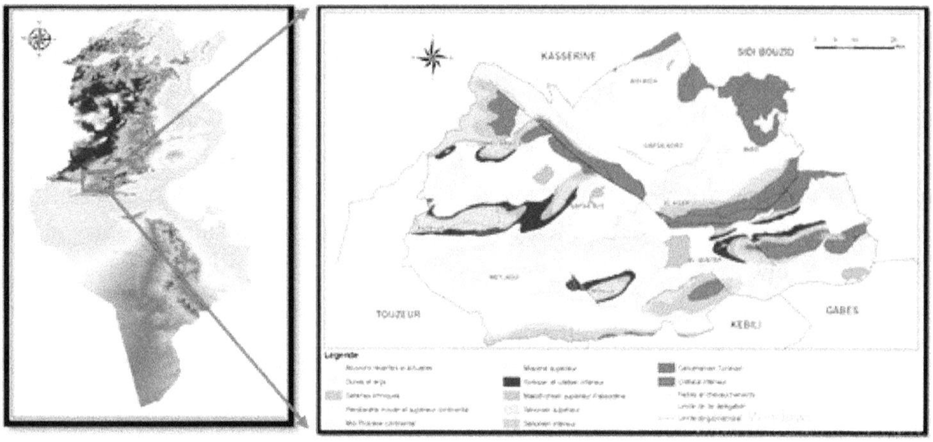

In terms of climatic conditions, no meteorological station was available in the Chott; the nearest station in Gafsa was used. The governorate of Gafsa occupies an intermediate position between the High Steppes north and the Saharan areas in the south. Therefore, it is characterized by the dominance of Mediterranean climate variations arid and semi-arid. Thus, the Chott Geuttar is dominated by an arid climate caused by high temperature and high thermal amplitude. The aridity of the climate explains the small amounts of rain falling in the governorate of Gafsa. Thus, rainfall rarely exceeds 200 mm per year.

METHODS AND MATERIALS

Field Expeditions and Core Sampling

The study of different systems was based on cores recuperated from the field. Indeed, coring is the collection of a sample from land or seabed. It helps to have a representative sample stratigraphically undisturbed. For the sedimentological study, 80 cm core was collected throughout the site. The coring was carried out using PVC tubes of 3 mm in thickness and 63 mm in diameter on an average depth of 1m. In the laboratory, the core was cut longitudinally with a grinding disc into two halves. The first half was the subject of a macroscopic lithological description (the composition, the description, the sediment sampling, the structure and the color of the sediments) and the second half, a granulometric, geochemical and geophysical studies. Once opened, cores should be sampled to save the geochemical signal. High-resolution sampling is useful to carry an efficient spectral analysis. Nonetheless, we are always in face of the financial handicaps which limit the number of studied samples. Also, after a certain threshold limit, increasing the sampling does not increase resolution. In our case, sampling was carried out each 1 cm. In doing so, 160 samples were obtained to study.

Colors Identification and Their Significance

Color is an important character of the sediment since it reflects its composition and chemical state. Recently, Marqueur et al. (2008) used the greyscale to study the high-resolution climatic variability in sebkha Mhabeul southeast Tunisia. Using sediment color to reconstruct El Nino activity during the Holocene, Rodbell et al. (1999) and Moy et al. (2002) also studied the lacustrine sediments from Laguna Pallcacocha located in Ecuador. Traditionally, sediment is described either as a Munsell code expressing Hue-Value-Chroma (HVC) or as a type of word-based (linguistic) description. The former is defined in the standard Geological Society of America (GSA) Rock Color Chart (Goddard et al., 1951), as 5GY 4/5 for green, for example. Unfortunately, the Munsell color system is not suitable for the combining of color terms, which is necessary to parse a multi-term description. It is psychometrically non-linear in Hue and Chroma but it is linear in Value. Despite the limitation of the color study discussed in Jenkins (2002), in this work, we will deal with color change within the core to investigate the cyclicity. The color study will be coupled with analyses to find any correlation between colors variability and geochemical analyses. Red, green, black, blue, yellow, magenta colors were measured from cores photos by Canvas software.

Geochemical Analysis

The geochemistry may be investigated in terms of elementary composition by studying the active ions (analogy with active carbonate, which is soluble in water) through a flame photometer.

Geophysical Exploration: Magnetic Susceptibility

Magnetic susceptibility is defined as the aptitude of a body to undergo magnetization when we applied a weak magnetic field. It characterizes the ratio of the induced magnetization when applying a low intensity of magnetic field and the field which creates it. Magnetic properties of soils are used in many domains as well as earth and environmental sciences (e.g., Blundell et al., 2009). Thus, it records sediment sources in fluvial, lacustrine, and marine sediments (Dearing, 2000), indicating atmospheric pollution (Hay et al., 1997). Also, it is considered a tool for archaeological mapping and prospecting (e.g., Dalan and Banerjee, 1996), as paleoclimate proxy in loess-paleo-sol sequences (Heller and Liu 1986; Maher and Thompson, 1995), as recorders of planetary atmospheric conditions (Barrón and Torrent, 2002; Retallack et al., 2003). Finally, the use of electromagnetic techniques helps to detect land mines (Hannam and Dearing, 2008). However, worldwide, the magnetic properties of minerals are still an enigma (Blundell et al., 2009).

Sedimentological Study: Grain Size Distribution

Sediment observation was carried out to see the diameter and the shape of the grain added to the other non-detritic present grains.

Spectral Analysis

As for data treatment, the simple spectral analysis and sinusoidal fitting were carried out through PAST software. PAST includes the Lomb periodogram technique for irregularly sampled data (Press et al., 1992) for spectrum analysis, using time values in the first column and dependent values in the second. The data are automatically detrended before analysis. The frequency and power value of the largest peak in the spectrum are exhibited, along with a probability that the peak might emerge from random data. The 0.01 and 0.05 significance levels are shown as dashed lines. To determine temporal climate cycles, it is necessary to date the time series.

RESULTS AND DISCUSSIONS

Colors Variability

Since it is simple and inexpensive, colors measurement is a promising tool for climatic investigation. To name but a few, many relevant studies such as Rodbell et al. (1999), Moy et al. (2002) and Marquer et al. (2008) dealt with colors to find out a well-established climatic cyclicity. The plot of colors variability along the core from Chott El Guettar (Figure 2) shows a change along with the core. This variability is due to a radical change in the geochemical and mineralogical compositions of sediment, which are intern due to changes in climatic conditions. Nonetheless, the straightforward interpretation remains difficult due to the interplay of geochemical and mineralogical factors in shaping the color. But the variability of colors is genetic changes in climatic conditions.

Figure 2. A Plot of colors variability along with the core from Chott El Guettar

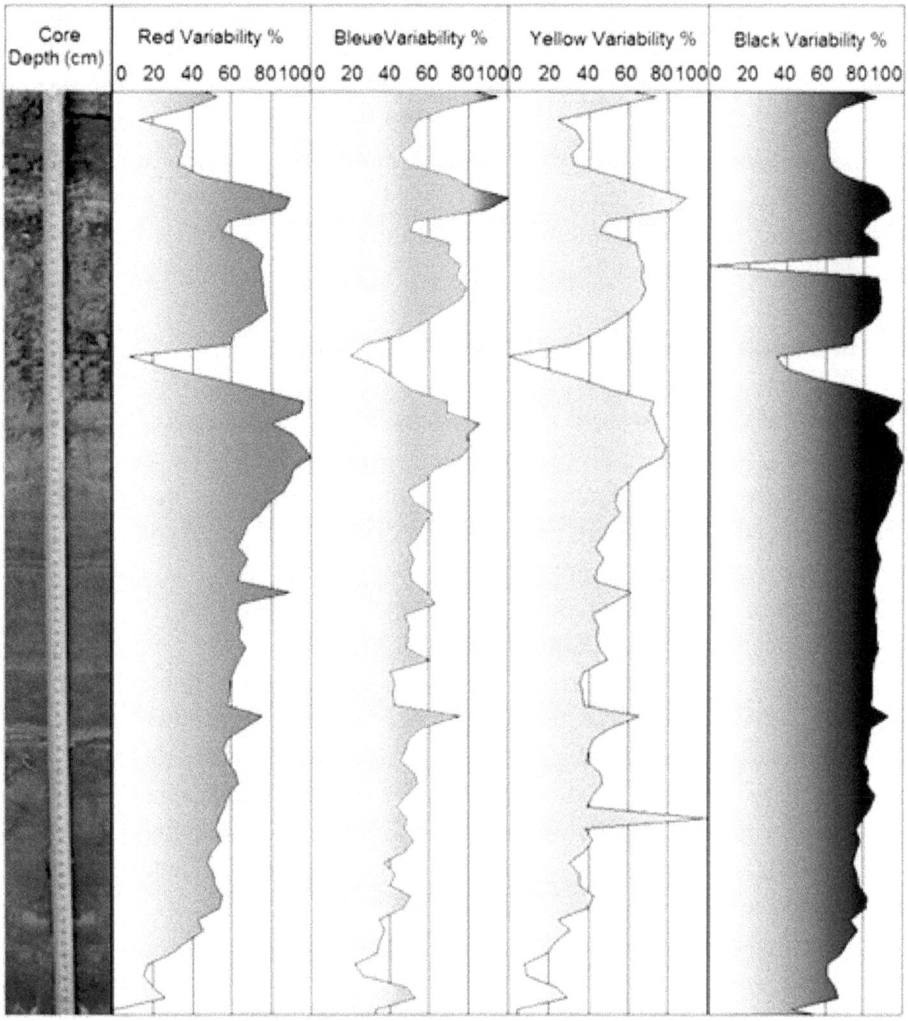

Spectral Analysis of Color Data

The spectral analysis of color data shows the individualization of the same cycles for all colors (Figures 3 and 4). A first cycle is centered around 1500 yr is well documented for the black and less pronounced for the blue (written in the figure blue). This cycle may be genetically related to the oceanography of the Atlantic Ocean. For instance, the Sea Surface Temperature record from the eastern Atlantic of NW Africa (Menocal et al., 2000) demonstrates a clear oceanic imprint with the 1500-year internal period well expressed during the Late Holocene. In addition, we notice the presence of a 1600 cycle for the red; some paleoclimatic records also show a cyclic period close to 1600 years with increasing intensity during the Late Holocene. Internal forcing, possibly oceanic driven, has been attributed to this second spectral signature, which is not apparent in solar activity records, as suggested by Broecker et al. (2001) and McManus et al. (1999). A third cycle is centered around 315 yr is well documented for the blue and the red and less pronounced for the black. This cycle may be attributed to solar activity. Dean (1997) found out through the geochemical record that the eolian activity exhibits distinct cyclicities with dominant periodicities of 400 and 84 yr.

Geochemical and Geophysical Variability

Evolution of Geochemical Elements and Magnetic Susceptibility

The study of the Chott El Geuttar saline system was based on the investigation of a core carried out in the northern part. In the laboratory, samples of core1 underwent the analysis of magnetic susceptibility, and chemical elements (Ca, Na K). Values of the magnetic susceptibility along the core show obvious cycles. The first stretch on 42 cm; it consists of two parts.

Due to the coming of ferromagnetic and ferrimagnetic minerals from the continental vicinity, the first section (30 cm) is characterized by a downward increase in magnetic susceptibility. The increase in magnetic susceptibility can be explained as a result of wet climatic conditions, which are the responsible agent for reaching the sebkha by an amount of detritic material via Wadi. This amount came from the Djbels, which are depressions or local base levels that receive sediment supplies and surround the sebkha system (Mefteh et al., 2015).

The second part ranging from 30 to 42 cm is characterized by low values of magnetic susceptibility, due to the deposition of diamagnetic mineral-like carbonate. As a result, a dry period is to account for this decrease, which causes the sedimentation of diamagnetic minerals. A second cycle extends all the way to the core's bottom. Recent works (Essefi et al., 2014; Mefteh et al., 2015) dealing with the spectral analysis of the sedimentary fillings of sebkha Mhabeul and sebkha Sidi Mansour show a cyclicity of 900 years. Hence, it may be said that these two cycles may extend to 1800 years. Since we have the depth and the age, we can estimate the rate of sedimentation as 0.45 mm/year. This rate is comparable to the rates of sedimentation at sebkhas Dkhila (Essefi et al., 2013), Souassi (Essefi et al., 2013), Mhabeul (Marquer et al., 2008; Essefi et al., 2014) and Sidi Mansour (Essefi et al., 2015). This millennial cyclicity was reported in the literature as the result of solar activity.

Figure 3. Spectral analysis of color (blue and red) data along with El Guettar core

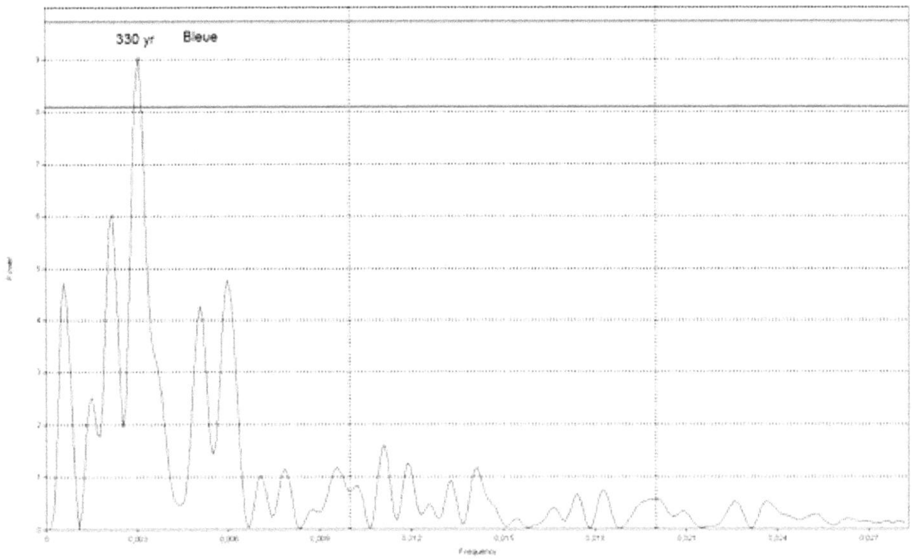

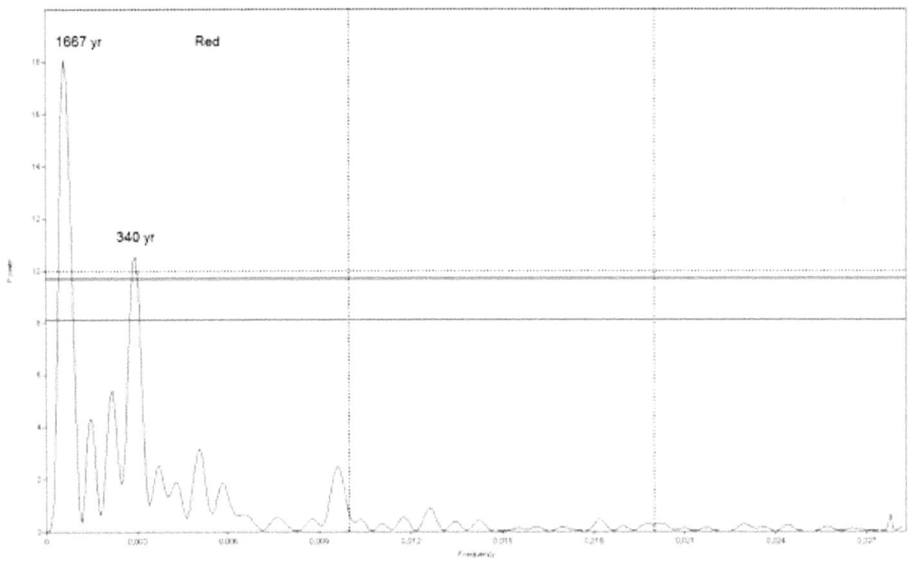

The geochemical study has touched on three chemical elements whose proportions significantly differed. Indeed, the results shown in the graphic below prove that sodium is the most representative element, whereas calcium and sodium exist in minor proportions. The active calcium rate shows a little variation as indicated and values range between 2 and 5 g/kg. The rate of active calcium is not only related to carbonate but it could also be probably linked to the existence of gypsum fraction

Figure 4. Spectral analysis of color (black) data along with El Guettar core

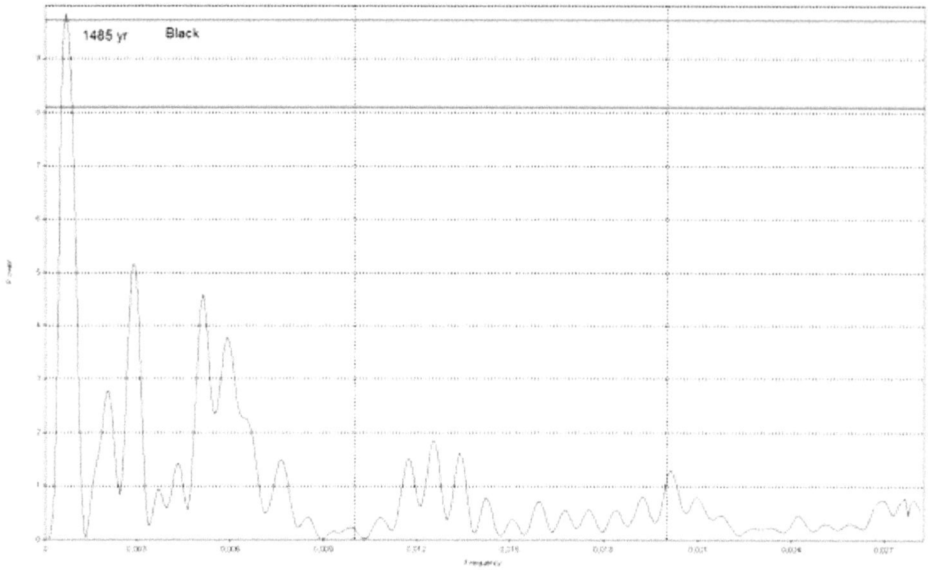

($CaSO_4$) (Figure 5). Along with the core, the proportion of potassium (K) presents low values that range from 0.2 to 1.5 g/Kg with slight fluctuation. This fraction is probably due to the existence of all. Along with the core, values of sodium content show an obvious oscillation.

Thus, the rate of sodium presents two intervals of decrease which stretch between 0-32cm and 49-75cm, characterized by an important amount of sodium with an average of 20g/Kg. Between these two intervals, we note a rise in the sodium proportion, and values exceed 20g/Kg. At the end of the cores, this interval is repeated as an indication of periodicity. We deduce then that the rate of Na had a close link with the precipitation of salt mineral (NaCl), thus, the periodicity shown in the graphic is probably due to changes in climatic conditions which explain the fluctuation of sodium rate. Subsequently, oscillation may results in the period of solar activity, indeed, in the arid period we distinguished precipitation of a high amount of (NaCl), while in the humid period, and a decrease in the rate of salt can be identified.

Spectral Analysis Based on the Geochemistry and the Magnetic Susceptibility

The cyclicity identified along the core may be found through the spectral analysis of data of the different parameters. The spectral analysis of data of and the amount of active Ca (Figure 6) and the magnetic susceptibility (Figure 7) shows the

Figure 5. Evolution of magnetic susceptibility and geochemical elements along with the core of the El Guettar saline system

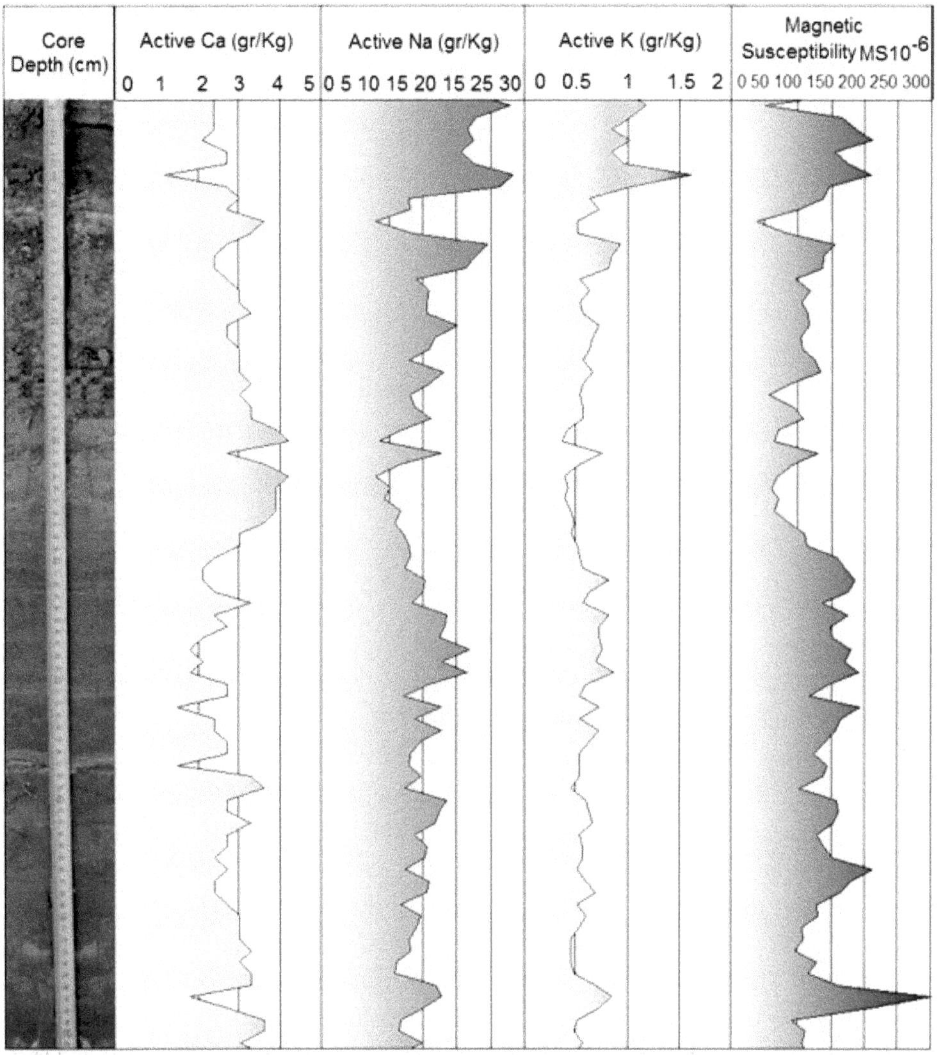

individualization of a cycle stretching along about 1000 yr within sebkhas Mhabeul (Essefi et al., 2014) and Sidi Mansour (Essefi et al., 2015), the about 1000 yr cycle were found out based on the magnetic susceptibility data. This also confirms the age-depth model used to estimate the dating at Chott El Guettar. The solar origin of this imprint is shown by the ^{10}Be record from the Greenland ice core (Vonmoos et al., 2006), and ^{14}C production rates (Bond et al., 2001). Moreover, Solanki and his collaborators demonstrate that the sunspot numbers and the residual ^{14}C give

Figure 6. Spectral Analysis based on the active Ca amount

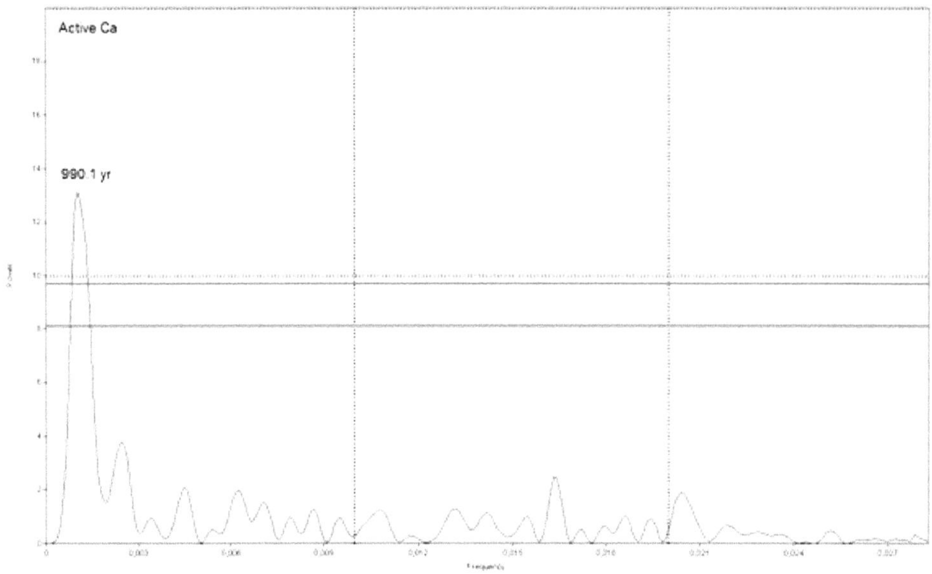

the same results (Solanki et al., 2004). The results of the numerous proxies reported here are used to identify the fingerprint in records from the Atlantic and Antarctic regions, as well as to confirm solar forcing.

On the other hand, the spectral analysis of data of the amounts of active Na and K shows the individualization of a cycle stretching along about 1200 yr (Figure 8). This cycle has never been discussed in the literature and its origin could by no means be discussed in this work. Nonetheless, errors during analyses or the variation of the rate of sedimentation would be the origin of the transformation of the 1000 yr to this hybrid cycle, which may only be mentioned with special care, unless it is a new finding. The SIP record (Nielsen et al., 2004) and SSIZ record (Crosta et al., 2007) reflect an internal force during the Late Holocene from 4000–5000 BP. But the frequencies are quite different with periods of 1250/1350 years in the Southern Atlantic Ocean instead of the ~1600 yr period in the North Atlantic. This study proves the link between saline systems in Tunisia and the global oceanography.

Sedimentary Analysis

Microscopic Observations of Samples

The study of the sedimentary filling of Chott El Guettar may be through the observations of sediments and the study of the grain size distribution. The observations

Figure 7. Spectral Analysis based on the magnetic susceptibility

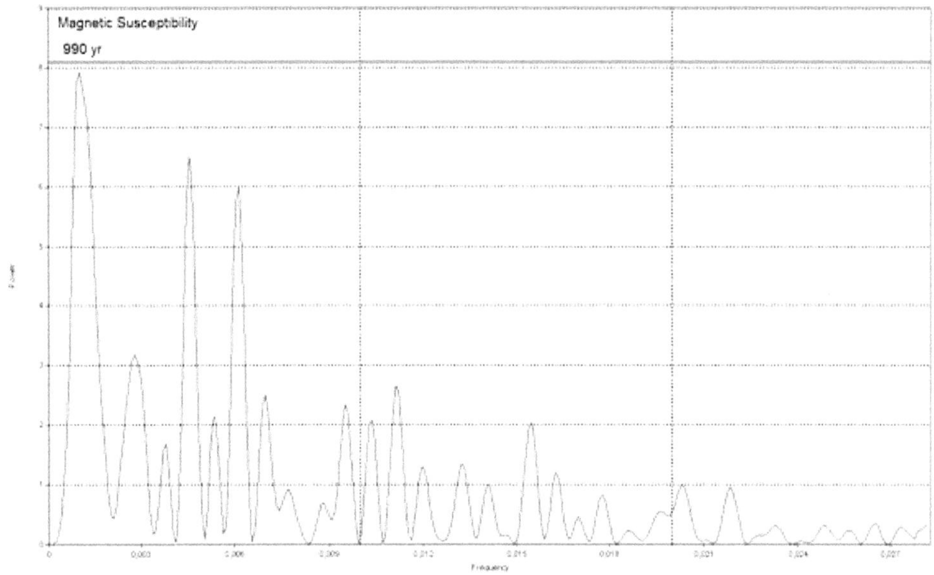

of four samples from the core of Chott El Guettar show a heterogeneous grain size distribution ranging from 10 µm to 400 µm (Figure 9). For example, the sample of 2 cm is made up of 10 µm to 200 µm grains; the shape is with sub-angles. On the other hand, the sample of 72 cm is made up of 40 µm to 400 µm round grains.

Evolution of Grain Size Parameters

The deposits of Tunisian saline systems are clayey. The fine grain size needs sophisticated technology. Fritsch apparatus allows the study of the grains ranging between 0.01 µm and 2 mm. Then, it produces a graph containing the cumulative and the frequency curves. Fritsch apparatus is equipped with sophisticated software (MaScontrol) providing the calculation of different statistical parameters. In addition to numerous graphs and other options, MaScontrol also offers its user statistical values, which may contain useful information for the evaluation of particle size distribution such as the mean, the mode, and the median. The plots of the mean, mode, median, kurtosis and skewness show the variability along with the core (Figure 10).

Figure 8. Spectral Analysis based on active K and Na amount

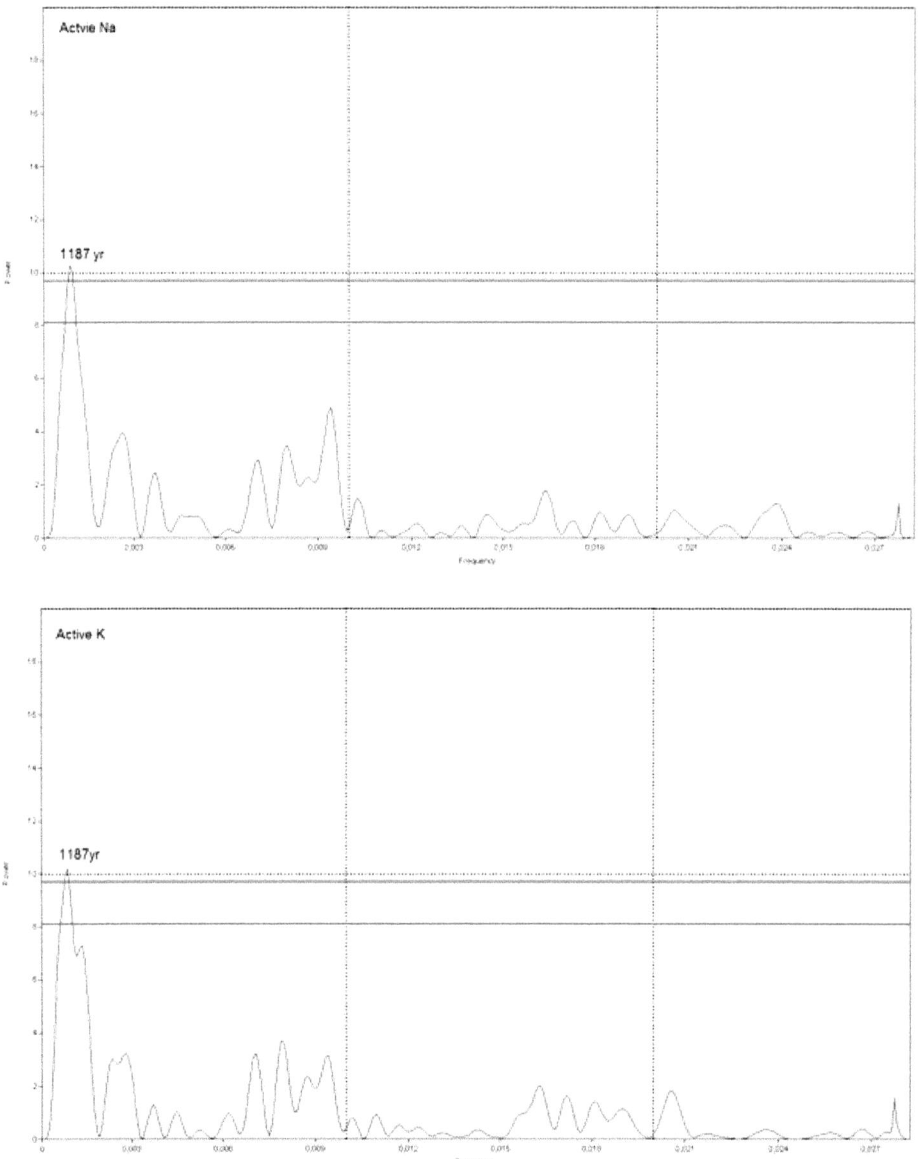

Figure 9. Microscopic observations of some samples from the core of Chott El Guettar

Spectral Analysis Based on Grain Size Parameters

The cyclicity identified along the core may be found through the spectral analysis of data of the grain size parameters. The spectral analysis of values of the mean grain size distribution shows the individualization of a cycle stretching along about 1000 yr (Figure11). The findings of the numerous proxies reported here are utilized to identify the fingerprint in records from the Atlantic and Antarctic regions, as well as to confirm solar forcing. The solar imprint is shown in the Norwegian ^{18}O record (Risebrobakken et al., 2003), which exhibits frequencies matching 1000–2500 year periods during the Early Holocene. In the South Atlantic Ocean, the sea ice record (Nielsen et al., 2004) presents solar periodicity identified with a continuous cycle of around 2500 years and with two others of around the 1000 and 1660 years only during the Early Holocene. It is worth noting that the 2500 cycles were not detected because the cycle is bigger than the core age. Also, we notice the setting of another cycle of 500 yr, which may be more detected through increasing the sampling resolution. Carcaillet et al. (2001) showed through the spectral analysis for the last 2000 years that the pollen-slide charcoal series shows a significant period of about 500 years.

Figure 10. Grain size parameters along with the core of Chott El Guettar

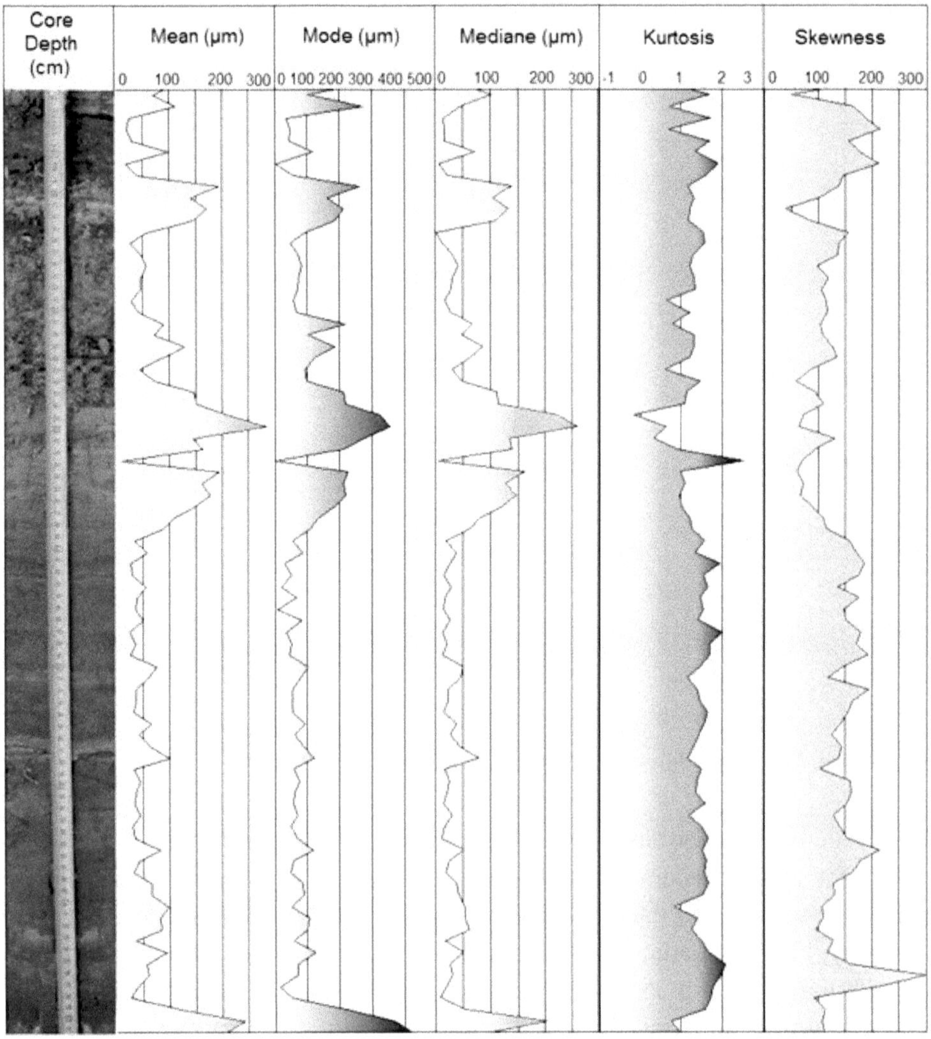

CONCLUSION

Holocene climate variability is different from that of the last glacial period and is the subject of intense research aimed at characterizing forcing factors of the observed climate oscillations evidenced in different regions of the world. This work on El Guettar allows the extraction of some millennial to decadal cycles. This Holocene cyclostratigraphy in inland systems is controlled by sun activity and also the oceanographic variability. The Late Holocene pattern proposed here confirms the importance of external forcing during the Holocene (solar activity:

Figure 11. Spectral Analysis based on grain size parameters along with the core of Chott El Guettar

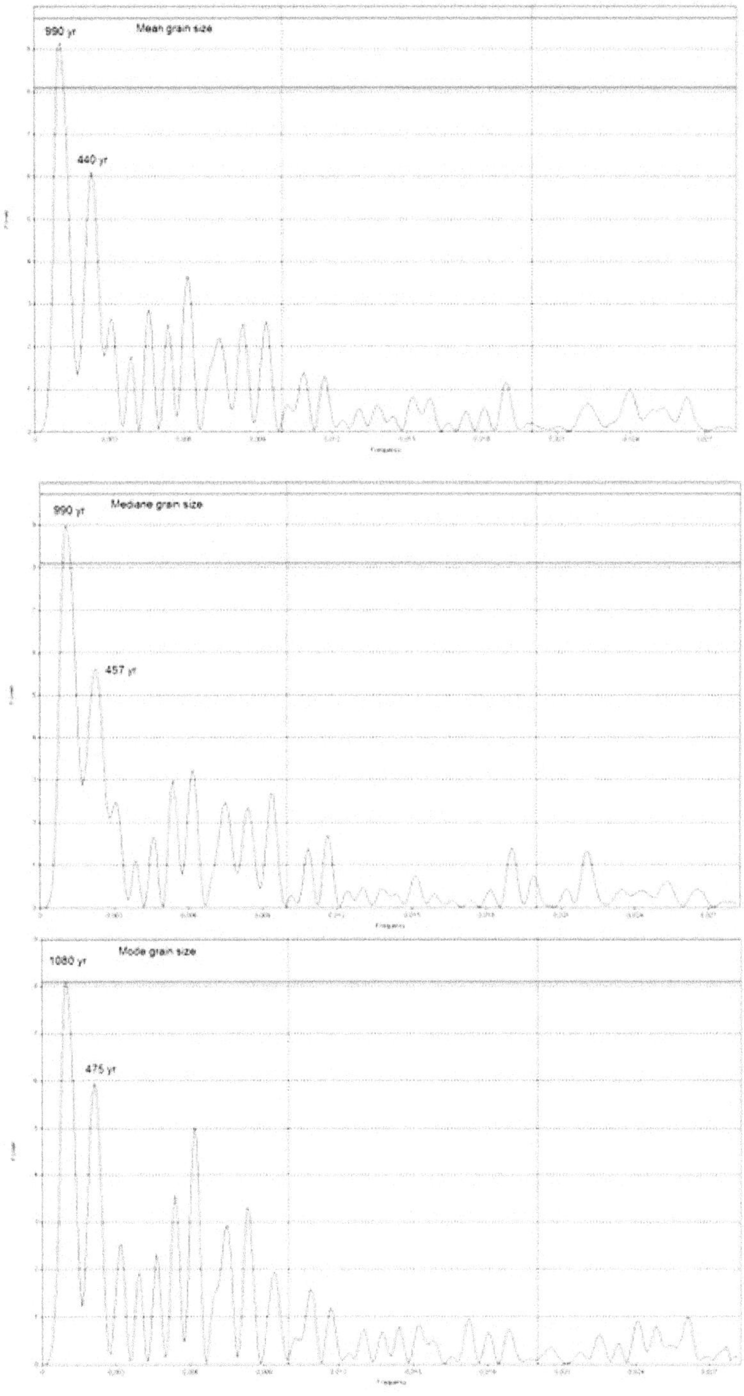

1000 years cyclicity and not the 2500 years during the entire Holocene), even if the signal is disturbed by meltwater fluxes. The Late Holocene is also marked by the gradual appearance of internal forcing (thermohaline circulation around 1500 years), associated with a stabilization of the signal. Also, centennial cycles were found (500 yr and 400 yr).

REFERENCES

Bedoui, S. (2016). *Multidisciplinary study of endorheic (El Guettar) and exoreic (El Melah) saline systems, Master*. Faculty of Sciences of Gabes-Tunisia.

Bedoui, S., Essefi, E., Besser, H., Benour, A., Ayadi, Y., Khelifi, F., & Hamed, Y. (2022). Geochemical, Mineralogical and Isotopic Fracturing during the Progressive Evaporation of Chott Djerid Brine (Southern Tunisia). In International Journal of Engineering Research in Africa (Vol. 58, pp. 107-125). Trans Tech Publications Ltd.

Bedoui, S., Essefi, E., Gharsalli, N., Wasli, A., & Ben Smida, N. (2017). *Multidisciplinary study of a core from chott el guettar, central Tunisia: Holocene cyclostratigraphy.* In The 1st International Symposium (WREIANA 2017) with oral communication March 24-25-26, 2017, Gafsa, Tunisia.

Ben Smida, N. (2016). Comparative study of exorheic (Boujmal and El Melah) and endorheic (Mchiguig and Mhabeul) saline systems: a geoeconomic approach. In *Institut Supérieur des Sciences et Techniques des Eaux de Gabès*. University of Gabes.

Blundell, A., Dearing, J. A., Boyle, J. F., & Hannam, J. A. (2009). Controlling factors for the spatial variability of soil magnetic susceptibility across England and Wales. *Earth-Science Reviews, 95*(3-4), 158–188. doi:10.1016/j.earscirev.2009.05.001

Broecker, W. S., Lynch-Stieglitz, J., Clark, E., Hajdas, I., & Bonani, G. (2001). What caused the atmosphere's CO2 content to rise during the last 8000 years? *Geochemistry Geophysics Geosystems, 2*(10), n/a. doi:10.1029/2001GC000177

Carcaillet, C., Bouvier, M., Fréchette, B., Larouche, A. C., & Richard, P. J. (2001). Comparison of pollen-slide and sieving methods in lacustrine charcoal analyses for local and regional fire history. *The Holocene, 11*(4), 467–476. doi:10.1191/095968301678302904

Crosta, X., & Koç, N. (2007). Chapter eight diatoms: From micropaleontology to isotope geochemistry. *Developments in Marine Geology, 1*, 327-369.

Dalan, R. A., & Banerjee, S. K. (1996). Soil magnetism, an approach for examining archaeological landscapes. *Geophysical Research Letters, 23*(2), 185–188. doi:10.1029/95GL03689

Dean, W. E. (1997). Rates, timing, and cyclicity of Holocene eolian activity in the north-central United States: Evidence from varved lake sediments. *Geology, 25*(4), 331–334. doi:10.1130/0091-7613(1997)025<0331:RTACOH>2.3.CO;2

Dearing, J. A. (1999). Holocene environmental change from magnetic proxies in lake sediments. *Quaternary Climates, Environments and Magnetism*, 231-278.

DeMenocal, P., Ortiz, J., Guilderson, T., & Sarnthein, M. (2000). Coherent high- and low-latitude climate variability during the Holocene warm period. *Science, 288*(5474), 2198–2202. doi:10.1126cience.288.5474.2198 PMID:10864866

Essefi, E. (2009). *Multidisciplinary study of Sidi El Hani Saline environment: the history and the climatic variability* [MS thesis]. Faculty of Sciences of Sfax, University of Sfax.

Essefi, E., Komatsu, G., Fairén, A. G., Chan, M. A., & Yaich, C. (2014). Groundwater influence on the aeolian sequence stratigraphy of the Mechertate–Chrita–Sidi El Hani system, Tunisian Sahel: Analogies to the wet-dry aeolian sequence stratigraphy at Meridiani Planum, Terby crater, and Gale crater, Mars. *Planetary and Space Science, 95*, 56–78. doi:10.1016/j.pss.2013.05.010

Essefi, E., Touir, J., Tagorti, M. A., & Yaich, C. (2013). Record of the climatic variability and the sedimentary dynamics during the last two millennia at sebkha Dkhila, eastern Tunisia. *International Scholarly Research Notices.*

Goddard, E. N., Trask, P. D., de Ford, R. K., Rove, O. N., Singewald, J. T., & Overbeck, R. M. (1951). *Rock Color Chart.* Geol. Soc.

Goodwin, E. J. (2017). Convention on Wetlands of International Importance, Especially as Waterfowl Habitat 1971 (Ramsar). In Elgar Encyclopedia of Environmental Law (pp. 101-108). Edward Elgar Publishing Limited.

Hamed, Y. (2011). The hydrogeochemical characterization of groundwater in Gafsa-Sidi Boubaker region (Southwestern Tunisia). *Arabian Journal of Geosciences.* Advance online publication. doi:10.100712517-011-0393-5

Hannam, J. A., & Dearing, J. A. (2008). Mapping soil magnetic properties in Bosnia and Herzegovina for landmine clearance operations. *Earth and Planetary Science Letters, 274*(3-4), 285–294. doi:10.1016/j.epsl.2008.05.006

Hay, K. L., Dearing, J. A., Baban, S. M. J., & Loveland, P. (1997). A preliminary attempt to identify atmospherically-derived pollution particles in English topsoils from magnetic susceptibility measurements. *Physics and Chemistry of the Earth, 22*(1-2), 207–210. doi:10.1016/S0079-1946(97)00104-3

Heller, F., & Tung-sheng, L. (1986). Palaeoclimatic and sedimentary history from magnetic susceptibility of loess in China. *Geophysical Research Letters, 13*(11), 1169–1172. doi:10.1029/GL013i011p01169

Jenkins, C. (2002). Automated digital mapping of geological color descriptions. *Geo-Marine Letters, 22*(4), 181–187. doi:10.100700367-002-0111-0

Lakhdar, A., Rabhi, M., Ghnaya, T., Montemurro, F., Jedidi, N., & Abdelly, C. (2009). Effectiveness of compost use in salt-affected soil. *Journal of Hazardous Materials, 171*(1-3), 29–37. doi:10.1016/j.jhazmat.2009.05.132 PMID:19576686

Maher, B. A., & Thompson, R. (1995). Paleorainfall reconstructions from pedogenic magnetic susceptibility variations in the Chinese loess and paleosols. *Quaternary Research, 44*(3), 383–391. doi:10.1006/qres.1995.1083

Marquer, L., Pomel, S., Abichou, A., Schulz, E., Kaniewski, D., & Van Campo, E. (2008). Late Holocene high-resolution palaeoclimatic reconstruction inferred from Sebkha Mhabeul, southeast Tunisia. *Quaternary Research, 70*(2), 240–250. doi:10.1016/j.yqres.2008.06.002

McManus, J. F., Oppo, D. W., & Cullen, J. L. (1999). A 0.5-million-year record of millennial-scale climate variability in the North Atlantic. *Science, 283*(5404), 971-975.

Mefteh, S., Essefi, E., Yaich, C., Jamoussi, F., & Medhioub, M. (2015). Correlation between magnetic susceptibility and mineral species along NWA-1 well, southern Tunisia: An overlap of the depositional environment, the climate, and the diagenesis. *Journal of African Earth Sciences, 103*, 89–101. doi:10.1016/j.jafrearsci.2014.12.007

Mokadem, N. (2012). Hydrodynamic and mineralization of deep groundwater in El Guettar–Gafsa basin ''SW Tunisia'' [Master]. Faculty of Sciences of Bizerte.

Mokadem, N., Hamed, Y., Hfaid, M., & Dhia, H. B. (2015). Hydrogeochemical and isotope evidence of groundwater evolution in El Guettar Oasis area, Southwest Tunisia. *Carbonates and Evaporites, 30*(4), 417–437. doi:10.100713146-015-0235-8

Moy, C. M., Seltzer, G. O., Rodbell, D. T., & Anderson, D. M. (2002). Variability of El Niño/Southern Oscillation activity at millennial timescales during the Holocene epoch. *Nature, 420*(6912), 162–165. doi:10.1038/nature01194 PMID:12432388

Nielsen, S. H., Koç, N., & Crosta, X. (2004). Holocene climate in the Atlantic sector of the Southern Ocean: Controlled by insolation or oceanic circulation? *Geology, 32*(4), 317–320. doi:10.1130/G20334.1

Retallack, G. J., Sheldon, N. D., Cogoini, M., & Elmore, R. D. (2003). Magnetic susceptibility of early Paleozoic and Precambrian paleosols. *Palaeogeography, Palaeoclimatology, Palaeoecology, 198*(3-4), 373–380. doi:10.1016/S0031-0182(03)00479-6

Risebrobakken, B., Jansen, E., Andersson, C., Mjelde, E., & Hevrøy, K. (2003). A high-resolution study of Holocene paleoclimatic and paleoceanographic changes in the Nordic Seas. *Paleoceanography, 18*(1), n/a. doi:10.1029/2002PA000764

Rodbell, D. T., Seltzer, G. O., Anderson, D. M., Abbott, M. B., Enfield, D. B., & Newman, J. H. (1999). A 15,000-year record of El Niño-driven alluviation in southwestern Ecuador. *Science, 283*(5401), 516–520. doi:10.1126cience.283.5401.516 PMID:9915694

Scholz, M., & Lee, B. H. (2005). Constructed wetlands: A review. *The International Journal of Environmental Studies, 62*(4), 421–447. doi:10.1080/00207230500119783

Smith, R. L. (1980). *Ecology and Field Biology* (3rd ed.). Harper and Row.

Solanki, S. K., Usoskin, I. G., Kromer, B., Schüssler, M., & Beer, J. (2004). Unusual activity of the Sun during recent decades compared to the previous 11,000 years. *Nature, 431*(7012), 1084–1087. doi:10.1038/nature02995 PMID:15510145

Tagorti, M. A., Essefi, E., Touir, J., Guellala, R., & Yaich, C. (2013). Geochemical controls of groundwaters upwelling in saline environments: Case study the discharge playa of Sidi El Hani (Sahel, Tunisia). *Journal of African Earth Sciences, 86*, 1–9. doi:10.1016/j.jafrearsci.2013.05.004

Torrent, J., & Barrón, V. (2002). Diffuse reflectance spectroscopy of iron oxides. Encyclopedia of Surface and Colloid Science, 1, 1438-1446.

KEY TERMS AND DEFINITIONS

Colors Identification: Used to study the high-resolution climatic variability in saline systems.

Coring Technique: The core is recuperated by percussion of a PVC tube of 3 mm thick and 63 mm in diameter.

Cyclostratigraphy: Is an integrated discipline that uses results from astronomy, climatology and various mathematical techniques.

Geochemistry: Application of chemical tools to understand geological phenomena.

High Resolution: This is used to find the climatic cycles.

Magnetic Susceptibility: Is defined as the aptitude of a body to undergo magnetisation when we applied a weak magnetic field.

Sebkha: Flat and salty bottom of a closed depression, without vegetation, characterized by saline efflorescence in dry periods, flooded by floods or upwelling of deep water in rainy periods.

Sedimentology: Is a branch of geology that studies the formation processes of sedimentary rocks.

Chapter 8
High Resolution Cyclostratigraphy During the Last Two Millennia Based on the Clayey Fraction Within the Mhabeul Wetland (Southeastern Tunisia)

Elhoucine Essefi
Faculty of Sciences of Sfax, University of Sfax, Tunisia

ABSTRACT

High-resolution proxy-based paleoenvironmental records derived from wetlands provide important insights into climate changes over centennial to millennial timescales. This work aims to study the climatic cyclicity of the clayey fraction along a 78 cm core from the saline system of Mhabeul, located in Southeastern Tunisia. Based on the age model of Essefi et al., the core covers the last two millennia (»2100 yr). A high-resolution sampling of 2mm was carried out to obtain 380 samples evenly distributed along the core. In terms of analyses, the cumulative curve of the grain size distribution carried out by Fritsch laser apparatus percentages of the Clayey fraction underwent the spectral analysis. Based on the spectral analysis, millennial cycles of » 2500 yr, and 1000 yr are related to solar forcing. Solar irradiation is most likely responsible for the cyclic characteristics at 500 yr (600 yr) frequency.

DOI: 10.4018/978-1-7998-9289-2.ch008

INTRODUCTION

Depending upon the culture, wetlands have been named differently (Essefi, 2009). But ambiguity and confusion are still common features in many languages. In the Arabic-speaking world, the word sebkha includes inland and coastal saline systems; its variant spellings such as salina, salare, salada, salar give approximately the same meaning. In many Spanish-speaking cultures, the word playa dominates in use. On the other hand, it is noticed that the English register proposes varieties of names to distinguish between these geological features. In spite of the ever present use of regional terms, several attempts for classification have been made to resolve confusion. To name but a few, playa, playa lake, lake playa, recharge playa, discharge playa, salt lake, pan, clay pan, saline lake, alkali flat, salt plain, dry lake, and salt flat are varieties of terminology frequently encountered in the literatures of the English-speaking cultures. These names could be given to a saline environment according to its geological features and the discipline of the study. A useful terminology of these saline environments was given by Rosen (1994). In this chapter, the words sebkha, endorheic basin and saline system are interchangeably used to indicate the the studied wetland of Mhabeul.

In order to predict the future climatic tendency, past climatic global changes and cyclicity should be inferred based on the sedimentary record of some depressions (e.g., oceans, seas) and wetlands (e.g., lakes and sebkhas). Due to their geographical locations, Mediterranean areas are considered as the most influenced region to past as well as future climate changes (IPCC, 2014, Jaouadi et al., 2016), associated with many factors such as polar ice retreat, anthropogenic warming and increased greenhouse gases concentrations (Giorgi, 2006; Fletcher and Zielhofer, 2013). The ecosystems of the Mediterranean region are particularly vulnerable to hydrological changes and strong demographic pressure (Lionello et al., 2006; IPCC, 2014; Jaouadi et al., 2016). In this vein, south-eastern Tunisia is currently suffering from increasing human pressure resulting in obvious pollution and arid climatic conditions, which reflect the degradation of its natural environment and the increasing desertification (Jaouadi et al., 2016; Essefi, 2021). The millennial and centennial scale variability of the Holocene climate in Tunisia (Essefi, 2021) is poorly understood, due to the lack of high-resolution records, especially of temperature.

The Holocene is an interglacial period that follows the last glacial of the Pleistocene. It is considered as a warm and stable interval, but now it is known to be characterized by pervasive climatic oscillations and abrupt climatic events (Mayewski et al., 2004, Wanner et al., 2008, Fletcher and Zielhofer, 2013), related to the Holocene cyclostratigraphy. Contrary to ancient cyclostratigraphy, which is controlled by the Milankovitch theory, the Holocene cyclostratigraphy is controlled by astronomical and oceanographic mechanisms. In fact, the millennial and decadal scale cycles of the Late Holocene sediment in saline systems (lakes and sebkhas) mentioned depending on geochemical and geophysical factors were less than likely explained by this theory (Essefi et al 2015a). Besides, short climatic cycles are explained by other astronomical and oceanographic causes (Essefi et al., 2014, 2015b, 2015c; Baumgarten and Wonik, 2015; Martinez-Ruiz et al., 2015). Then, Sun activity and its effect on climatic cyclicity are the most popular (Essefi et al., 2015a). The geographical position of sebkha Mhabeul (the study area), situated at the southeastern side of the Mediterranean Sea and in North Africa, has a consequence on the interpretation of the factors intervening in the climatic variations. Indeed, from the climatic and hydrological viewpoint, the Mediterranean Sea is depending on the Atlantic Ocean. So, in addition to the solar activity, the Holocene cyclostratigraphy is probably related to the North Atlantic Oscillation (NAO) and the Mediterranean Oscillation (MO).

Many climatic cycles related to astronomic and/or oceanographic forcing are worldwide recorded in depressions, namely oceans, seas and wetlands. The about 2000yr and 5000yr cycles and the about 2500yr and 1000 yr cycles (Debret et al., 2007) may be related to solar activity. Also, the about 500 yr cycle can be probably linked to the sun's activity. The 1500yr climate cycle is linked to solar activity and the atmospheric dynamics and their relation to the ocean's circulation (Bond et al., 2001). Also, Jiménez-Espejo et al. (2014) indicated that the about1500-1400 yr and 2600-2200 yr cycles are probably related to the solar activity and the NAO circulation. According to Jian et al. (2000) and Debret et al. (2007), the 1500 yr and 700-800 yr cycles are possibly related to the oceanic thermohaline circulation. The about 1600 yr cycle (Debret et al., 2007) and the 1700 yr cycle (Essefi et al., 2015a) are related to the oceanic activity. Minor secondary millennial to centennial cycles of 650 yr, 1087 yr, and 3000 yr are not well recognized (Essefi et al., 2015a). The basis of this rhythm throughout the Holocene is still controversial. Bond et al. (1997, 2001) linked the Holocene period to solar radiation. However, due to the rare occurrence of the corresponding period of months in the anti-solar reconstructions, the Holocene time variation mechanism has caused widespread controversy (Essefi, 2021). Paleoclimates based mainly on proxies have increasingly confirmed the existence of Holocene climate cycles from one century to one thousand years (for example, Obrochta et al., 2012; Swindles et al., 2012; Ojala et al., 2015; Stebich et al., 2015; Xu et al., 2014, 2019, 2020).

In this work, the aim was to study the Holocene cyclostratigraphy along with a core from an endorheic system (sebkha Mhabeul) located at the coast of the Mediterranean Sea. This study is based on the cyclicity of the grain size distribution along the78 cm core covering the last two millennia.

STUDY AREA

The sebkha of Mhabeul (southeastern of Tunisia) is located in the southwestern of the Mediterranean Sea, belonging to the Medenine governorate (Figure 1). This region is characterized by an arid climate. According to the temporal variation of the precipitation, there are two episodes: dry (summer) and wet (autumn, winter and spring). The total rainfall average is about 174.5 mm/yr with major monthly variations. Indeed, the dry summer is marked by high temperatures reaching sometimes 40 °C in the shade and by drought due to the almost total absence of precipitation. However, the wet period is characterized by low temperature and high rainfall. During the period 1985-2015, the maximum monthly evaporation reaches its maximum in July and August (average 215-225mm) and the low monthly evaporation is recorded in December and January (average 100 -105 mm). The winds of the wet period, coming from the direction North-West, are hot and dry. The winds of the dry period, which are rather from the Sahara, are coming from the West and South-West directions. The latter corresponds to the hot Sirocco winds blowing in summer and spring. The sebkha Mhabeul is covered with marine and lacustrine Quaternary deposits and limited to the south by an escarpment of Mesozoic limestones. Precipitation and runoff from small hydrographic systems are the major factors that essentially influence the sedimentation mode of this sebkha.

METHODS AND MATERIALS

To set the current climatic conditions in southern Tunisia, the index of precipitation in the region of Medenine was studied during the last century (1909-1996) in relation tosolar activity (Sunspot Number (SSN), Western Mediterranean Oscillation Index (WeMedOi), and the Sothern Oscillation Index (SOi). These parameters were plotted together and the Principal Component Analysis PCA was carried out to set a link of correlation and/causality between them. During the field mission of January 2015, a 78 cm core was collected from the sebkha of Mhabeul, just near the core of Essefi et al. (2014). The coring was carried out using PVC tubes of 63 mm diameter and 3.2 mm in thickness on an average depth of 1m. In the laboratory, the core was longitudinally cut with a grinding disc into two halves. A high-resolution sampling of 2 mm was carried out to obtain 396 samples. In this paper, we focused on the evolution of the clayey fraction, mean, median and mode of the grain size distribution. The FRITSCH apparatus measuring the grain size distribution automatically calculated the parameters. It should be noted that particles size is used as a proxy to interpret climate changes and to determine the sea-level fluctuation (Hartmann, 2007). At the sebkha Mhabeul, the decrease of the particle size and the increased clay amount

Figure 1. The sebkha of Mhabeul within the Mediterranean and Atlantic contexts: (a) Surface Atlantic water circulation in the Mediterranean basin (Bianchi et al., 2012); (b) Geographical location of sebkha Mhabeul (Southeastern Tunisia): (c) Cites of cores of previous and in-progress works on the sebkha of Mhabeul

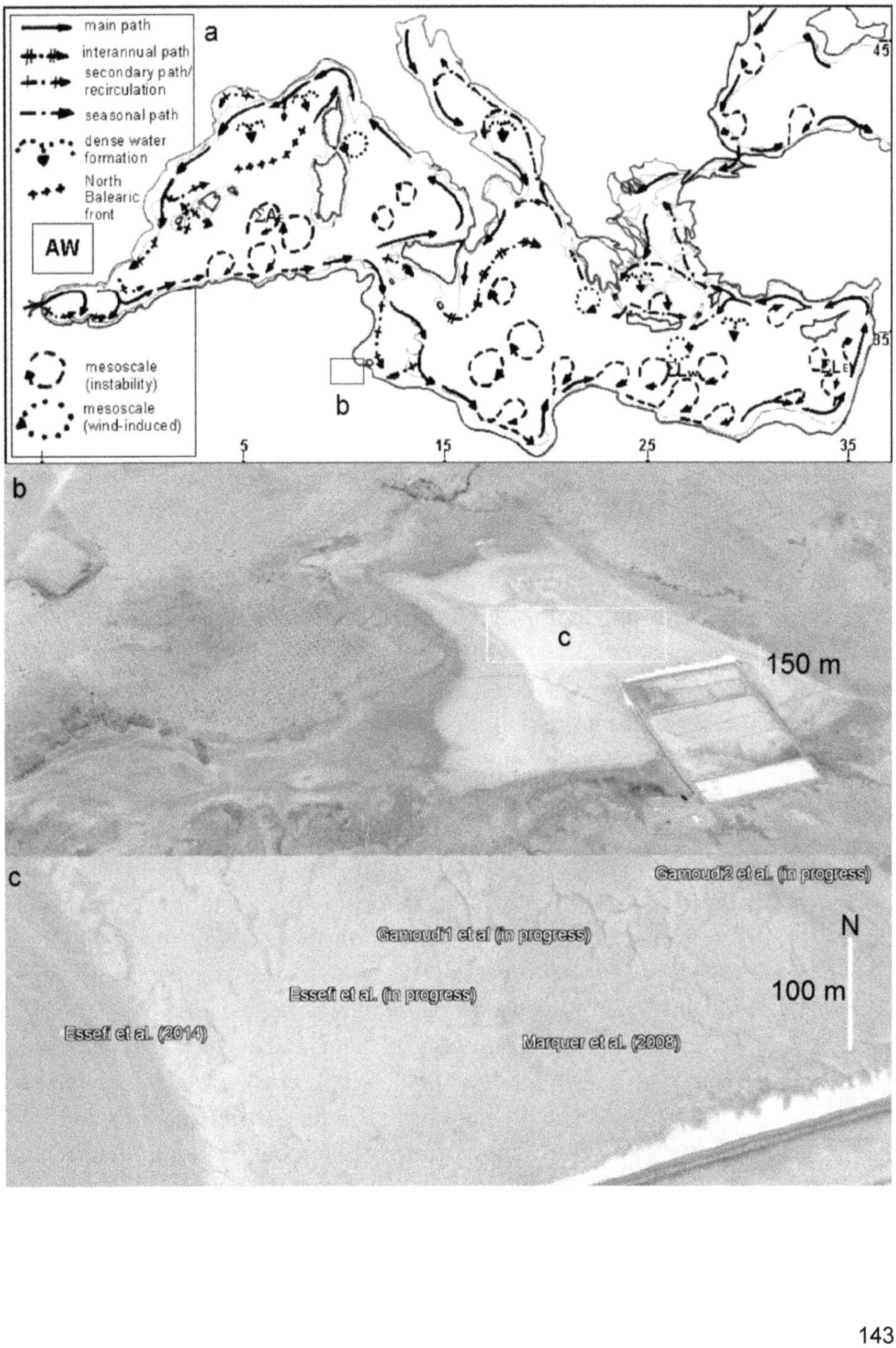

is an indication of flood events, i.e., humid phases (Gammoudi et al., 2019; Essefi, 2021). Thespectral analysis was carried out through PAST software. PAST includes the Lomb periodogram algorithm for unevenly sampled data. This software uses as input, in the first column, the time values given and in the second column, the dependent values. As result, the highest peak in the spectrum is presented with its frequency and power value. Due to the proximity (Figure 1) of our core (78 cm) and the core of Essefi et al. (2014) (Figure 1c), the age model of Marquer et al. (2008) was used.

RESULTS AND DISCUSSION

Current Climate in Southern Tunisia and the Relation with Astronomical, Oceanographic, and Atmospheric Forcing

The Western Mediterranean Oscillation (WeMO or WeMOi) is an index measuring the difference between the normalized atmospheric pressure recorded at Padua (45.40°N, 11.48°E) in northern Italy, and at San Fernando, Cadiz (36.28°N, 6.12°W) in southwestern Spain. While Padua is an area with relatively high barometric variability due to the influence of the Central European High, San Fernando is often under the influence of the Upper Azores. This index was proposed as a teleconnection by researchers from the climatology group of the University of Barcelona, offering an alternative to the better known NAO to study the variability of precipitation in eastern Spain, in regions like Catalonia, Valencia and Murcia. Some climatologists believe that the WeMOi barometric model is causally related to, and therefore partially predictive, the variability of precipitation on the eastern side of the Iberian Peninsula. A positive phase of the WeMOi generally shows high pressure in the Gulf of Cadiz region and an area of low pressure at the edge of the Ligurian Sea, while a negative phase of the WeMOi will show a low pressure in the Gulf of Cadiz and a high-pressure area in central Europe. During the positive phase, the prevailing winds in the Iberian Peninsula are typically west and north-west, originating in the North Atlantic region; these winds, as they reached the eastern side of the peninsula, swept through the mainland areas of the peninsula, so they became dry and hot (westerly winds) or cooler but just as dry (northwest). On the other hand, a negative WeMOi phase is associated with humid air flows which have crossed the Mediterranean Sea; these are therefore loaded with humidity when they reach the east side of the Iberian Peninsula, which leads to an increase in precipitation - sometimes torrential - in this area. In relation to the climate in southern Tunisia, a negative phase of WeMOi is accompanied by a rainy phase in southern Tunisia (Figure 1a). A fact, the PCA Principal Component Analysis shows a negative correlation between the rainfall index and WeMOi (Figure 1b). The Southern Oscillation Index (SOI) is a standardized index based on the pressure differences observed at sea level between Tahiti and Darwin, Australia. SOI is a measure of the large-scale fluctuations in atmospheric pressure that occur between the western and eastern tropical Pacific (i.e., the state of the Southern Oscillation) during El Niño and La Niña episodes. In general, the smoothed SOI time series correspond very well to changes in ocean temperatures in the eastern tropical Pacific. The negative phase of SOI is lower than the normal air pressure in Tahiti and higher than the normal air pressure in Darwin. Prolonged periods of negative (positive) SOI values coincide with abnormally warm (cold) ocean waters in the eastern tropical Pacific, typical of El Niño (La Niña) episodes. In relation to the climate in southern Tunisia, a positive phase of SOI is accompanied by a rainy phase in southern Tunisia (Figure 2). A fact, the PCA Principal Component Analysis shows a positive correlation between rainfall index and SOI (Figure 2b). Scientist's follow solar cycles by counting sunspots - cold areas the size of a planet from the Sun where strong magnetic loops cross the visible surface of the star. Recently, Gil-Alana and Yaya (2014) compared the record for the number of sunspots and the record for temperature and found that they differ: the record for the number of sunspots is characterized by a dominant cycle of 11 years while the record for temperature seems to be characterized by a "singularity" or "pole" in the spectral density function at the "zero" frequency. In relation to the climate in southern Tunisia, a decrease in the number of sunspots is accompanied by a rainy phase in southern Tunisia (Figure 2a). However, the PCA Principal Component Analysis does not show this positive correlation between the rainfall index and the number of sunspots (Figure 2b).

Age Model and Extrapolation

Dating in continental sebkha remains a major handicap to set a precise age model, which is interns the principal requirement not only to infer the record of the paleoclimatic variability but also to find out the Holocene cyclostratigraphy. Fortunately, dating was already carried out by Marquer et al. (2008) and Essefi et al. (2014). During field expeditions, our core was meant to be close (78 m) to these cores (Figure 1c). This condition allowed us to calculate the rate of sedimentation at every depth and to extrapolate the age of the remaining part (Figure 3). Hence, the rate of sedimentation during the Late Holocene at sebkha Mhabeul was in the range of 3.116 by the year and the extrapolated age at the bottom of the core is 2100 yr. This model proves vital to pinpoint climatic events and find out high frequency cyclicity.

Figure 2. (a) Evolution of rain index in southern Tunisia, WeMOi, SOI, SSN, from 1910 to 1995 and (b) their Principal Component Analysis

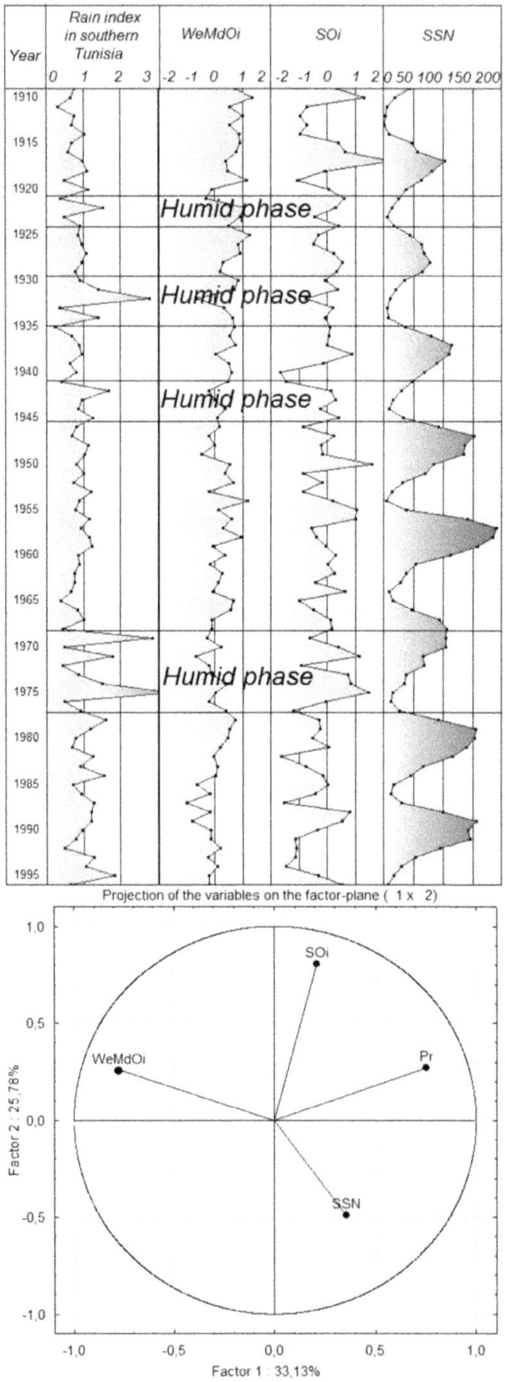

Figure 3. Age model and extrapolation based on tephrochronological dating of Essefi et al. (2014)

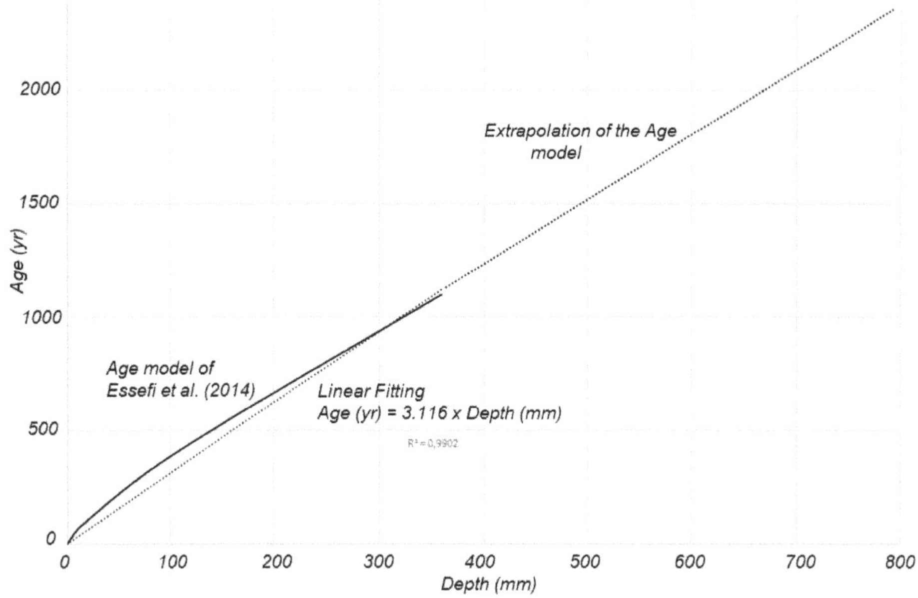

Variability of the Grain Size Distribution During the Last Two Millennia

To infer the climatic signal within the sedimentary filling of sebkha Mhabeul, we may rely on the study of the grain size distribution, specifically the variability of the clayey fraction and mean, mode and median of the grain size distribution. Previous granulometric study of the clay pan of Mhabeul (Essefi et al., 2015b; Gammoudi et al., 2019) showed the importance of coarse fractions to highlight changes related to the sedimentary dynamics. In this vein, we consider the fine fraction completing the total grain may also prove vital to infer the climatic conditions. Indeed, the percentages of different sandy and silty fractions contribute to the understanding of climatic conditions and the depositional environment. As it is shown in Figure 4, the sedimentary filling of the studied sebkha is basically dominated by the very fine sand and the silt. With the size rise of the sandy fraction the percentages decrease. Thus, rates of very fine sand along the core are very high unlike the percentages of coarse sand which vary between 0% and 7%. The coarse sand only appears in some locations along the core. Generally, the high amount of very fine sand is related to wet climatic conditions. On the other hand, the appearance of the coarse sand

Figure 4. Evolution of the sand and silt fractions percentages along the core of sebkha Mhabeul

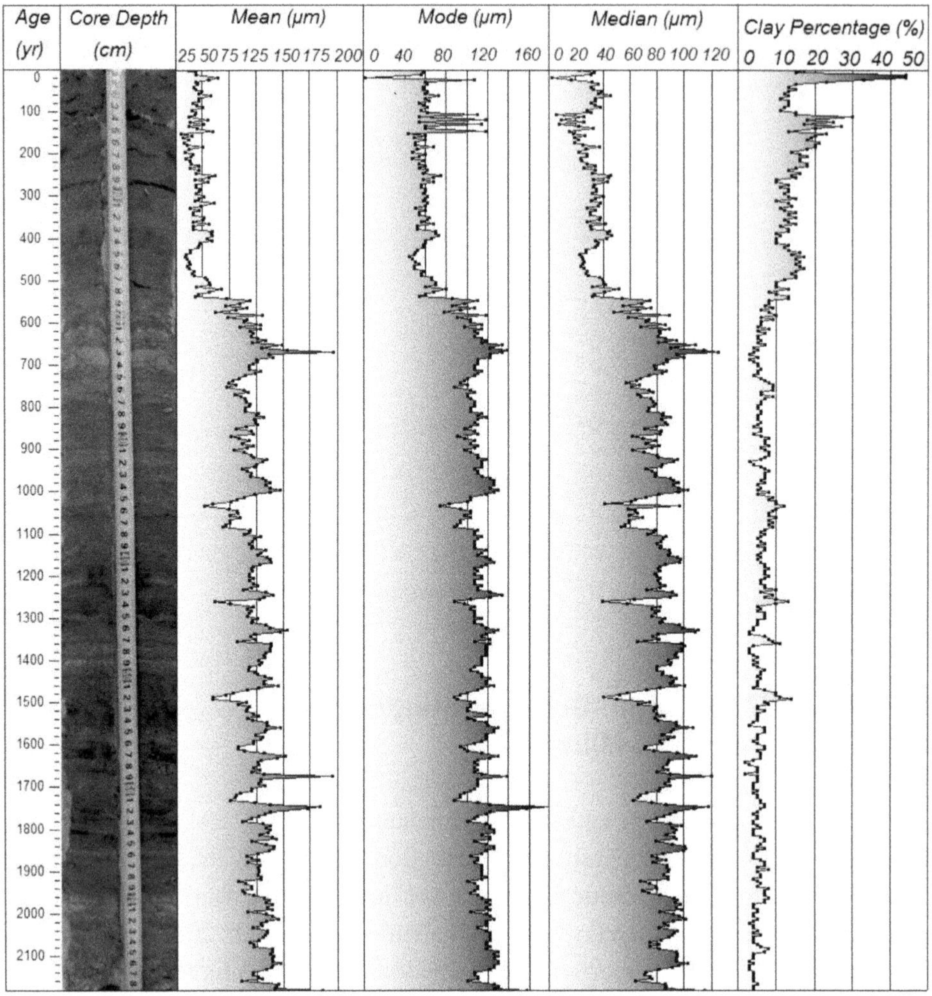

fraction is related to arid climatic conditions. However, fractions below 0.063 mm have high percentages along the core. That is an indicator of a humid climate.

Climatic Cycles and Factors Involved in Holocene Cyclostratigraphy

So, to define the main factors intervening in the climatic changes at sebkha Mhabeul (Southwestern Mediterranean) during the last two millennia, spectral analysis was used. The spectral analyses of values of mode and mean show the individualization

Figure 5. Spectral analysis based on the sand fractions percentages: Mode (a); mean (b)

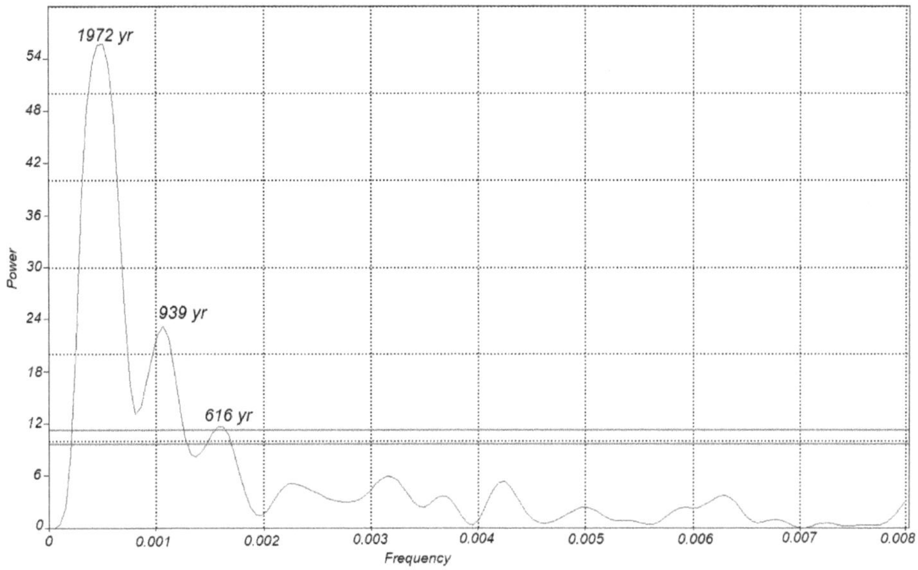

of many climatic cycles (Figure 5). First, the spectral analysis of both mode and mean values shows the individualization of cycles stretching along about 2000 yr, 1000 yr, and 600yr. Indeed, the cycle of 2000yrand 1000 yr may be related to the solar activity (Rodrigo-Gámiz et al., 2014).

The spectral analyses of values of the percentages of sand fractions show the individualization of many climatic cycles (Figure 6). First, the spectral analysis of values of coarse sand shows the individualization of cycles stretching along about 2500 yr, 1000yr, and 600 yr. These cycles were attributed to the climatic solar forcing (Li et al., 2021; Huang et al., 2020; Warken et al., 2020; Weber et al., 2020).

Global Record of the Holocene Cyclostratigraphy

As for the current evolution of climate in southeastern Tunisia is linked to astronomical, oceanographic and atmospheric conditions. The study area is one of the southwestern Mediterranean regions. In fact, there is a probably link between the West Mediterranean aridification and the temperature depressions in the North Atlantic region (Bond et al., 2001; Faust et al., 2004). Moreover, Bond et al. (2001) highlight a North Atlantic cooling with cyclicity of »1450 years for the Holocene (Bond events) (Faust et al., 2004). Indeed, the comparison between the Madjerda sedimentation rates (Faust et al., 2004) and the Hematite curve (Bond et al., 2001)

Figure 6. Spectral analysis based on the sand fractions percentages: clay (a); median (b)

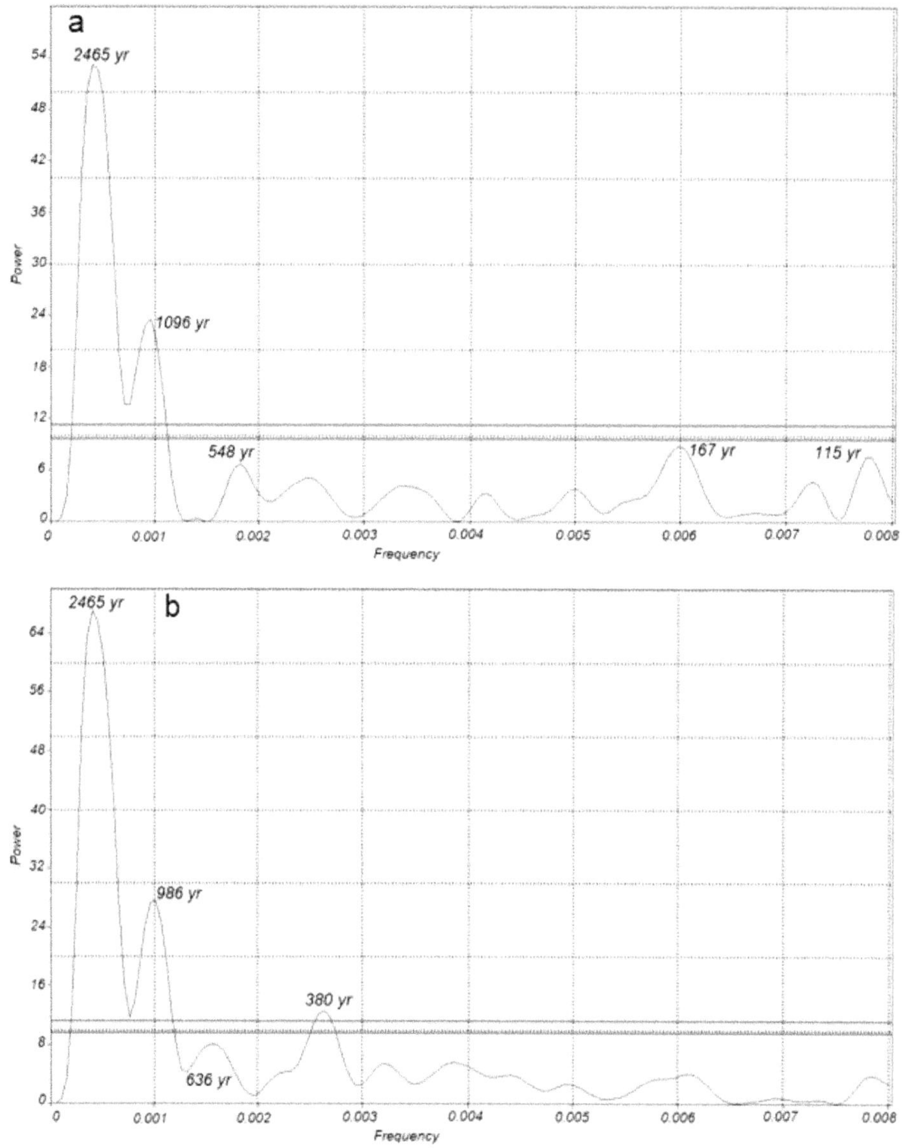

shows a good agreement between the North Atlantic cooling and the increased alluviation in the Mediterranean region (Tunisia)during the late Holocene. This comparison makes the supposition of more arid conditions with higher fluvial dynamics around 1600yr and 400 yr, corresponding to the Bond events (Faust et al., 2004). Therefore, the late Holocene is marked by a linkage between the Mediterranean basin and the North Atlantic region (Faust et al., 2004). In fact, the saline environments in Tunisia, depending on its geographical position, are affected by the global oceanography of the Mediterranean Sea, which is related to the Atlantic Oscillation. As result, the Holocene cyclostratigraphy within the saline system of Mhabeul depends not only on the solar activity but also probably depends on the North Atlantic Oscillation (NAO), the Mediterranean Oscillation (MO). The regional circulation pattern (MO), which is derived from the Sea Level Pressure data across the extended Mediterranean region, is associated with the cyclogenesis activity in the Mediterranean, essentially in the bay of Genoa. Atmospheric circulation is important to the hydrological budget of the Mediterranean Sea (Jalali et al., 2016). The NAO is the dominant mode of atmospheric variability at mid-latitudes in the North Atlantic region. The NAO index is an indicator of atmospheric circulation and weather patterns. Indeed, when the index is positive, Europe and the eastern US are mild and wet while Greenland and northern Canada are cold and dry. Besides, when the index is negative, the southerly position of NAO results in enhanced winter rainfall over Southern Europe and the Northwestern Mediterranean Sea (Jalali et al., 2016). De La Torre et al. (2006) supposed that solar variations arising on timescales of years can have an impact on the climatic conditions by affecting major modes of atmospheric variability. Actually, the NAO covers the Northern Hemisphere during solar maximum phases and, it is confined to the eastern Atlantic sector and the troposphere during minimum solar phases (De La Torre et al., 2006).

CONCLUSION

The study of the sebkha El Mhabeul through the investigation of the grain size distribution data visualizes the presence of wet and arid climate conditions. The treatment of the results of the sedimentological study shows some significant cycles. Indeed, the spectral analysis of values of the mode and mean shows the individualization of cycles stretching along about 2000 yr, 1000 yr, and 600yr. Indeed, the cycles of 2000 yr and 500 yr are probably related to solar activity. The 1700 yr cycle is related to the oceanic activity. Though, the rest of the cycles of 630yr, » 433-383 yr and 122 yr are probably related to other oceanographic and atmospheric factors. As a result, the Holocene cyclostratigraphy is the result of the combination of solar activity, oceanographic and atmospheric parameters.

REFERENCES

Baumgarten, H., & Wonik, T. (2015). Cyclostratigraphic studies of sediments from Lake Van (Turkey) based on their uranium contents obtained from downhole logging and paleoclimatic implications. *International Journal of Earth Sciences, 104*(6), 1639–1654. doi:10.100700531-014-1082-x

Bond, G., Kromer, B., Beer, J., Muscheler, R., Evans, M. N., Showers, W., Hoffmann, S., Lotti-Bond, R., Hajdas, I., & Bonani, G. (2001). Persistent solar influence on North Atlantic climate during the Holocene. *Science, 294*(5549), 2130–2136. doi:10.1126cience.1065680 PMID:11739949

Bond, G., Showers, W., Cheseby, M., Lotti, R., Almasi, P., DeMenocal, P., ... & Bonani, G. (1997). A pervasive millennial-scale cycle in North Atlantic Holocene and glacial climates. *Science, 278*(5341), 1257-1266.

De La Torre, L., Gimeno, L., Añel, J. A., & Nieto, R. (2007). The role of the solar cycle in the relationship between the North Atlantic Oscillation and Northern Hemisphere surface temperatures. *Advances in Atmospheric Sciences, 24*(2), 191–198. doi:10.100700376-007-0191-x

Debret, M., Bout-Roumazeilles, V., Grousset, F., Desmet, M., McManus, J. F., Massei, N., & Trentesaux, A. (2007). *The origin of the 1500-Essefi, E. (2021). Apport des Approches Géochimique et Sédimentologique au Fonctionnement des Systèmes Endoréiques et Exoréiques de l'Holocène à l'Anthropocène: Intérêts Environnemental et Géoéconomique* [Contribution of Geochemical and Sedimentological Approaches to the Functioning of Endorheic and Exoreic Systems from the Holocene to the Anthropocene: Environmental and Geoeconomic Interests]. Institut Supérieur des Sciences et Techniques des Eaux de Gabès.

Essefi, E. (2009). *Multidisciplinary study of Sidi El Hani Saline Environment: The Geological History and the Climatic Variability*. Academic Press.

Essefi, E., Gharsalli, N., Kalabi, S., Ben Ameur, M., & Yaich, C. (2015d). Spectral analysis of a core from the Sebkha of Sidi Mansour, southern Tunisia: The Holocene cyclostratigraphy. *J Geophys Remote Sens, 4*, 141. doi:10.4172/2169-0049.1000141

Essefi, E., Gharsalli, N., Salhi, S., & Yaich, C. (2015b). Cyclicity of grain size parameters along a Core from the Gleysol of Sebkha Ennoual, southeastern Tunisia. *J Basic Appl Res Int, 15*(5–3), 124–134.

Essefi, E., Gharsalli, N., & Yaich, C. (2015a). Climatic significance aeolian sedimentary filling of sebkha Mhabeul, southeastern Tunisia: Contribution of the genetic approach. *J Basic Appl Res Int, 15*(2–4), 140–152.

Essefi, E., Gharsalli, N., & Yaich, C. (2015c). Geophysical and geochemical study of the silico-evaporitic sedimentary filling of Boujmal wetland, eastern Tunisia: Inferring the climatic signal within groundwater noise. *J Basic Appl Res Int, 15*(2–1), 14–25.

Essefi, E., Mefteh, S., Medhioub, M., & Yaich, C. (2014). Magnetic study of the heated and unheated sedimentary fillings of Sebkha Mhabeul, Southeast Tunisia: A geophysical method for paleoclimatic investigation and tephrochronological dating. *International Journal of Geophysics, 2014*, 2014. doi:10.1155/2014/908395

Faust, D., Zielhofer, C., Escudero, R. B., & del Olmo, F. D. (2004). High-resolution fluvial record of late Holocene geomorphic change in northern Tunisia: Climatic or human impact? *Quaternary Science Reviews, 23*(16-17), 1757–1775. doi:10.1016/j.quascirev.2004.02.007

Fletcher, W. J., & Zielhofer, C. (2013). Fragility of Western Mediterranean landscapes during Holocene rapid climate changes. *Catena, 103*, 16–29. doi:10.1016/j.catena.2011.05.001

Gammoudi, A., Essefi, E., & Rigane, H. (2019). Variability of the grain-sized fractions during the last three millennia within the sedimentary fillings of the saline system of Mhabeul: The Holocene cyclostratigraphy. *Arabian Journal of Geosciences, 12*(9), 1–11. doi:10.100712517-019-4472-3

Gil-Alana, L. A., & Yaya, O. S. (2014). The relationship between oil prices and the Nigerian stock market. An analysis based on fractional integration and cointegration. *Energy Economics, 46*, 328–333. doi:10.1016/j.eneco.2014.10.001

Giorgi, F. (2006). Climate change hot-spots. *Geophysical Research Letters, 33*(8), L08707. doi:10.1029/2006GL025734

Giraudeau, J., Cremer, M., Manthé, S., Labeyrie, L., & Bond, G. (2000). Coccolith evidence for instabilities in surface circulation south of Iceland during Holocene times. *Earth and Planetary Science Letters, 179*(2), 257–268. doi:10.1016/S0012-821X(00)00113-8

Hartmann, D. (2007). From reality to model: Operationalism and the value chain of particle-size analysis of natural sediments. *Sedimentary Geology, 202*(3), 383–401. doi:10.1016/j.sedgeo.2007.03.013

Huang, C., Rao, Z., Li, Y., Yang, W., Liu, L., Zhang, X., & Wu, Y. (2020). Holocene summer temperature in arid central Asia linked to millennial-scale North Atlantic climate events and driven by centennial-scale solar activity. *Palaeogeography, Palaeoclimatology, Palaeoecology, 556*, 109880. doi:10.1016/j.palaeo.2020.109880

IPCC: Climate Change. (2014). *Synthesis Report. Contribution of working groups I. II and III to the fifth assessment report of the intergovernmental panel on climate change*. IPCC.

Jalali, B., Sicre, M. A., Bassetti, M. A., & Kallel, N. (2016). Holocene climate variability in the north-western Mediterranean Sea (gulf of lions). *Climate of the Past*, *12*(1), 91–101. doi:10.5194/cp-12-91-2016

Jaouadi, S., Lebreton, V., Bout-Roumazeilles, V., Siani, G., Lakhdar, R., Boussoffara, R., Dezileau, L., Kallel, N., Mannai-Tayech, B., & Combourieu-Nebout, N. (2016). Environmental changes, climate and anthropogenic impact in south-east Tunisia during the last 8 kyr. *Climate of the Past*, *12*(6), 1339–1359. doi:10.5194/cp-12-1339-2016

Jian, Z., Wang, P., Saito, Y., Wang, J., Pflaumann, U., Oba, T., & Cheng, X. (2000). Holocene variability of the Kuroshio current in the Okinawa Trough, northwestern Pacific Ocean. *Earth and Planetary Science Letters*, *184*(1), 305–319. doi:10.1016/S0012-821X(00)00321-6

Jiménez-Espejo, F. J., García-Alix, A., Jiménez-Moreno, G., Rodrigo-Gámiz, M., Anderson, R. S., Rodríguez-Tovar, F. J., Martínez-Ruiz, F., Giralt, S., Delgado Huertas, A., & Pardo-Igúzquiza, E. (2014). Saharan aeolian input and effective humidity variations over western Europe during the Holocene from a high altitude record. *Chemical Geology*, *374*, 1–12. doi:10.1016/j.chemgeo.2014.03.001

Li, N., Sharifi, A., Chambers, F. M., Ge, Y., Dubois, N., Gao, G., Li, D., Liu, L., Liu, H., Wang, J., Niu, H., Meng, M., Liu, Y., Zhang, G., & Jie, D. (2021). Linking Holocene East Asian monsoon variability to solar forcing and ENSO activity: Multi-proxy evidence from a peatland in Northeastern China. *The Holocene*, *31*(6), 966–982. doi:10.1177/0959683621994662

Lionello, P., Malanotte-Rizzoli, P., Boscolo, R., Alpert, P., Artale, V., Li, L., ... Xoplaki, E. (2006). The Mediterranean climate: an overview of the main characteristics and issues. *Developments in Earth and Environmental Sciences, 4*, 1-26.

Marquer, L., Pomel, S., Abichou, A., Schulz, E., Kaniewski, D., & Van Campo, E. (2008). Late Holocene high resolution palaeoclimatic reconstruction inferred from Sebkha Mhabeul, southeast Tunisia. *Quaternary Research*, *70*(2), 240–250. doi:10.1016/j.yqres.2008.06.002

Martinez-Ruiz, F., Kastner, M., Gallego-Torres, D., Rodrigo-Gámiz, M., Nieto-Moreno, V., & Ortega-Huertas, M. (2015). Paleoclimate and paleoceanography over the past 20,000 yr in the Mediterranean Sea Basins as indicated by sediment elemental proxies. *Quaternary Science Reviews*, *107*, 25–46. doi:10.1016/j.quascirev.2014.09.018

Mayewski, P. A., Rohling, E. E., Stager, J. C., Karlén, W., Maasch, K. A., Meeker, L. D., ... Steig, E. J. (2004). Holocene climate variability. *Quaternary Research, 62*(3), 243–255. doi:10.1016/j.yqres.2004.07.001

Obrochta, S. P., Miyahara, H., Yokoyama, Y., & Crowley, T. J. (2012). A re-examination of evidence for the North Atlantic "1500-year cycle" at Site 609. *Quaternary Science Reviews, 55*, 23–33. doi:10.1016/j.quascirev.2012.08.008

Ojala, A. E. K., Launonen, I., Holmström, L., & Tiljander, M. (2015). Effects of solar forcing and North Atlantic oscillation on the climate of continental Scandinavia during the Holocene. *Quaternary Science Reviews, 112*, 153–171. doi:10.1016/j.quascirev.2015.01.021

Rosen, M. R. (1994). *The importance of groundwater in playas: A review of playa classifications and the sedimentology and hydrology of playas*. Academic Press.

Stebich, M., Rehfeld, K., Schluetz, F., Tarasov, P. E., Liu, J., & Mingram, J. (2015). Holocene vegetation and climate dynamics of NE China based on the pollen record from Sihailongwan Maar Lake. *Quaternary Science Reviews, 124*, 275–289. doi:10.1016/j.quascirev.2015.07.021

Swindles, G. T., Morris, P. J., Baird, A. J., Blaauw, M., & Plunkett, G. (2012). Ecohydrological feedbacks confound peat-based climate reconstructions. *Geophysical Research Letters, 39*(11), n/a. doi:10.1029/2012GL051500

Wanner, H., Beer, J., Bütikofer, J., Crowley, T. J., Cubasch, U., Flückiger, J., Goosse, H., Grosjean, M., Joos, F., Kaplan, J. O., Küttel, M., Müller, S. A., Prentice, I. C., Solomina, O., Stocker, T. F., Tarasov, P., Wagner, M., & Widmann, M. (2008). Mid-to Late Holocene climate change: An overview. *Quaternary Science Reviews, 27*(19-20), 1791–1828. doi:10.1016/j.quascirev.2008.06.013

Warken, S. F., Schorndorf, N., Stinnesbeck, W., Hennhoefer, D., Stinnesbeck, S. R., Förstel, J., Steidle, S. D., Avilés Olguin, J., & Frank, N. (2021). Solar forcing of early Holocene droughts on the Yucatán peninsula. *Scientific Reports, 11*(1), 1–12. doi:10.103841598-021-93417-z PMID:34230539

Weber, N., Lazar, B., Gavrieli, I., Yechieli, Y., & Stein, M. (2021). Gypsum deltas at the Holocene Dead Sea linked to grand solar minima. *Geophysical Research Letters, 48*(6), e2020GL091034.

Xu, D., Lu, H., Chu, G., Liu, L., Shen, C., Li, F., Wang, C., & Wu, N. (2019). Synchronous 500-year oscillations of monsoon climate and human activity in Northeast Asia. *Nature Communications, 10*(1), 1–10. doi:10.103841467-019-12138-0 PMID:31511523

Xu, D., Lu, H., Chu, G., Shen, C., Li, F., Wu, J., Wang, L., Li, H., Yu, Y., Jin, Y., & Wu, N. (2020). Asynchronous 500-year summer monsoon rainfall cycles between Northeast and Central China during the Holocene. *Global and Planetary Change*, *195*, 103324. doi:10.1016/j.gloplacha.2020.103324

Xu, D., Lu, H., Wu, N., Liu, Z., Li, T., Shen, C., & Wang, L. (2013). Asynchronous marine-terrestrial signals of the last deglacial warming in East Asia associated with low-and high-latitude climate changes. *Proceedings of the National Academy of Sciences of the United States of America*, *110*(24), 9657–9662. doi:10.1073/pnas.1300025110 PMID:23720306

Compilation of References

ADB. (2020). *Climate risk country profile: Sri Lanka.* Asian Development Bank.

Akgüneş, G. (2020, June 21). *We must learn to live with blowfish.* Milliyet. https://www.milliyet.com.tr/yazarlar/gurkan-akgunes/balon-baligiyla-yasamayi-ogrenmeliyiz-6239709

Aksu, A. E., Konuk, T., Ulug, A., & Duman, M., & PIPER, D. W. (1990). Quaternary tectonic and sedimentary history of Eastern Aegean Sea shelf area. *Jeofizik*, *4*(1), 3–35.

Albay, T. (2015, July 30). *Fish is declining in Turkey.* AA. https://www.aa.com.tr/tr/turkiye/turkiyede-balik-azaliyor/21
379

Alexander, P., Rounsevell, M. D., Dislich, C., Dodson, J. R., Engström, K., & Moran, D. (2015). Drivers for global agricultural land use change: The nexus of diet, population, yield and bioenergy. *Global Environmental Change*, *35*, 138–147. doi:10.1016/j.gloenvcha.2015.08.011

Arslan, D., Çiçek, K., Döndüren, Ö., & Ernoul, L. (2021). Threat Ranking to Improve Conservation Planning: An Example from the Gediz Delta, Turkey. *Land (Basel)*, *10*(12), 1381. doi:10.3390/land10121381

Arslan, D., Oliver, A., Yaşar, Ç., Ismail, İ. B., Döndüren, Ö., Ernoul, L., ... Çiçek, K. (2018). Distribution and current status of herpetofauna in the Gediz Delta (Western Anatolia, Turkey). *Herpetology Notes*, *11*, 1–15.

Ataol, M., & Onmuş, O. (2021). Wetland loss in Turkey over a hundred years: Implications for conservation and management. *Ecosystem Health and Sustainability*, *7*(1), 1930587. doi:10.1080/20964129.2021.1930587

Avdan, A. O. (2020). *The making of consurbia: Conservation, urbanization, and socio-environmental change in Turkey's Gediz Delta.* [Thesis, Simon Fraser University, Canada].

Bambaradeniya, C. (2003). Freshwater wetlands in Sri Lanka: their conservation significance and current status. *National symposium on wetland conservation and management* (pp. 19-24). Colombo7: IUCN.

Baumgarten, H., & Wonik, T. (2015). Cyclostratigraphic studies of sediments from Lake Van (Turkey) based on their uranium contents obtained from downhole logging and paleoclimatic implications. *International Journal of Earth Sciences*, *104*(6), 1639–1654. doi:10.100700531-014-1082-x

Bax, N., Williamson, A., Aguero, M., Gonzalez, E., & Geeves, W. (2003). Marine invasive alien species: A threat to global biodiversity. *Marine Policy*, *27*(4), 313–323. doi:10.1016/S0308-597X(03)00041-1

Bedir, M. (1989). *Geodynamics of the sedimentary bassins of Sahel Mahdia (eastern Tunisia), from Aptian to present: sismo-stratigraphy, sismo-tectonics, and sismo-structural* [PhD Thesis]. University of El Manar Tunis.

Bedoui, S., Essefi, E., Besser, H., Benour, A., Ayadi, Y., Khelifi, F., & Hamed, Y. (2022). Geochemical, Mineralogical and Isotopic Fracturing during the Progressive Evaporation of Chott Djerid Brine (Southern Tunisia). In International Journal of Engineering Research in Africa (Vol. 58, pp. 107-125). Trans Tech Publications Ltd.

Bedoui, S., Essefi, E., Gharsalli, N., Wasli, A., & Ben Smida, N. (2017). *Multidisciplinary study of a core from chott el guettar, central Tunisia: Holocene cyclostratigraphy.* In The 1st International Symposium (WREIANA 2017) with oral communication March 24-25-26, 2017, Gafsa, Tunisia.

Bedoui, S. (2016). *Multidisciplinary study of endorheic (El Guettar) and exoreic (El Melah) saline systems, Master*. Faculty of Sciences of Gabes-Tunisia.

Ben Smida, N. (2016). Comparative study of exorheic (Boujmal and El Melah) and endorheic (Mchiguig and Mhabeul) saline systems: a geoeconomic approach. In *Institut Supérieur des Sciences et Techniques des Eaux de Gabès*. University of Gabes.

Bilecenoğlu, M., & Çınar, M. E. (2021). Alien Species Threat across Marine Protected Areas of Turkey—An Updated Inventory. *Journal of Marine Science and Engineering*, *9*(10), 1077. doi:10.3390/jmse9101077

Blundell, A., Dearing, J. A., Boyle, J. F., & Hannam, J. A. (2009). Controlling factors for the spatial variability of soil magnetic susceptibility across England and Wales. *Earth-Science Reviews*, *95*(3-4), 158–188. doi:10.1016/j.earscirev.2009.05.001

Bolca, M., Özen, F., & Güneş, A. (2014). Land use changes in Gediz Delta (Turkey) and their negative impacts on wetland habitats. *Journal of Coastal Research*, *30*, 756–764.

Bond, G., Showers, W., Cheseby, M., Lotti, R., Almasi, P., DeMenocal, P., ... & Bonani, G. (1997). A pervasive millennial-scale cycle in North Atlantic Holocene and glacial climates. *Science*, *278*(5341), 1257-1266.

Bond, G., Kromer, B., Beer, J., Muscheler, R., Evans, M. N., Showers, W., Hoffmann, S., Lotti-Bond, R., Hajdas, I., & Bonani, G. (2001). Persistent solar influence on North Atlantic climate during the Holocene. *Science*, *294*(5549), 2130–2136. doi:10.1126cience.1065680 PMID:11739949

Bostancı, S. H., & Yıldırım, S. (2022). The Role of Municipalities in Achieving Water Security: The Case of Turkey. In R. Castanho (Ed.), *Handbook of Research on Sustainable Development Goals, Climate Change, and Digitalization* (pp. 268–286). IGI Global. doi:10.4018/978-1-7998-8482-8.ch017

Brinkman, R. (2021). Impact of climatic change on coastal agriculture. *Climatic Change*, 235–245.

Broecker, W. S., Lynch-Stieglitz, J., Clark, E., Hajdas, I., & Bonani, G. (2001). What caused the atmosphere's CO2 content to rise during the last 8000 years? *Geochemistry Geophysics Geosystems*, *2*(10), n/a. doi:10.1029/2001GC000177

Çalkaya, M. (2020, December 3). *Fishermen welcomed the promotion of blowfish fishing*. AA. https://www.aa.com.tr/tr/ekonomi/balikcilar-balon-baligi-avciliginin-tesvik-edilmesini-olumlu-karsiladi/2064002

Carcaillet, C., Bouvier, M., Fréchette, B., Larouche, A. C., & Richard, P. J. (2001). Comparison of pollen-slide and sieving methods in lacustrine charcoal analyses for local and regional fire history. *The Holocene*, *11*(4), 467–476. doi:10.1191/095968301678302904

Carol, E., Pilar Alvarez, M. D., Santucci, L., Candanedo, I., & Arcia, M. (2022). Origin and dynamics of surface water-groundwater flows that sustain the Matusagaratí Wetland, Panamá. *Aquatic Sciences*, *84*(1), 1–15. doi:10.100700027-021-00847-y

CBI. (2021, August 30). *What is the demand for fish and seafood on the European market?* CBI [the Centre for the Promotion of Imports from developing countries]. https://www.cbi.eu/market-information/fish-seafood/what-demand

CEA. (1993). *Wetland site report and conservation management plan*. CEA.

Cherkauer, D. S., & Nader, D. C. (1989). Distribution of groundwater seepage to large surface water bodies: The effect of hydraulic heterogeneities. *Journal of Hydrology (Amsterdam)*, *109*(1-2), 151–165. doi:10.1016/0022-1694(89)90012-7

Chillasse, L., & Dakki, M. (2004). Potentialités et statuts de conservation des zones humides du Moyen-Atlas (Maroc), avec référence aux influences de la sécheresse. [Potentialities and conservation status of wetlands in the Middle Atlas (Morocco), with reference to the influences of drought.] *Science et changements planétaires/Sécheresse, 15*(4), 337-345.

Christoffersen, L.E. (1997). IUCN: A Bridge-Builder for Nature Conservation. *Green Globe Yearbook*, 59–70

Çinar, M. E., Bilecenoğlu, M., Yokeş, M. B., Öztürk, B., Taşkin, E., Bakir, K., Doğan, A., & Açik, Ş. (2021). Current status (as of end of 2020) of marine alien species in Turkey. *PLoS One*, *16*(5), e0251086. doi:10.1371/journal.pone.0251086 PMID:33945562

Constanza, R., d'Arge, R., DeGroot, R., Farbe, S., Grasso, M., Hannon, B., & O'Neill, R. V. (1997). The value of the worlds ecosystem services and natural capital. *Nature*, *387*(6630), 253–260. doi:10.1038/387253a0

Crandell, J. (2021). Geographic Distribution of the wetlands. *Britannica*. https://www.britannica.com/ science/wetland/Geographic-distribution-of-wetlands

CRIP. (2018). *Strategic Environmental Assessment of Development of River Basin Level Flood and Drought Mitigation Plans.* Kotte: Consulting Engineers & Architects Associated (Pvt.) Ltd. http://crip.lk/wp-content/uploads/2019/01/Kelani-River-SEA-Study-Final-Report.pdf

Crosta, X., & Koç, N. (2007). Chapter eight diatoms: From micropaleontology to isotope geochemistry. *Developments in Marine Geology, 1*, 327-369.

Dalan, R. A., & Banerjee, S. K. (1996). Soil magnetism, an approach for examining archaeological landscapes. *Geophysical Research Letters, 23*(2), 185–188. doi:10.1029/95GL03689

Dammalage, T., & Jayasinghe, N. (2019). Land-Use Change and Its Impact on Urban Flooding: A Case Study on Colombo District Flood on May 2016. *Engineering, Technology &. Applied Scientific Research, 9*(2), 3887–3891.

Davidson, N. C. (2014). How much wetland has the world lost? Long-term and recent trends in global wetland area. *Marine and Freshwater Research, 65*(10), 934–941. doi:10.1071/MF14173

Davidson, N. C., Fluet-Chouinard, E., & Finlayson, C. M. (2018). Global extent and distribution of wetlands: Trends and issues. *Marine and Freshwater Research, 69*(4), 620–627. doi:10.1071/MF17019

Dayathilaka, D., Lokupitiya, E., & Wijerathne, V. (2020). Estimation of aboveground and belowground carbon stocks in urban freshwater wetlands of Sri Lanka. *Carbon Balance and Management, 15*(17), 17. doi:10.118613021-020-00152-5 PMID:32876789

De La Torre, L., Gimeno, L., Añel, J. A., & Nieto, R. (2007). The role of the solar cycle in the relationship between the North Atlantic Oscillation and Northern Hemisphere surface temperatures. *Advances in Atmospheric Sciences, 24*(2), 191–198. doi:10.100700376-007-0191-x

Dean, W. E. (1997). Rates, timing, and cyclicity of Holocene eolian activity in the north-central United States: Evidence from varved lake sediments. *Geology, 25*(4), 331–334. doi:10.1130/0091-7613(1997)025<0331:RTACOH>2.3.CO;2

Dearing, J. A. (1999). Holocene environmental change from magnetic proxies in lake sediments. *Quaternary Climates, Environments and Magnetism*, 231-278.

Debret, M., Bout-Roumazeilles, V., Grousset, F., Desmet, M., McManus, J. F., Massei, N., & Trentesaux, A. (2007). *The origin of the 1500-Essefi, E. (2021). Apport des Approches Géochimique et Sédimentologique au Fonctionnement des Systèmes Endoréiques et Exoréiques de l'Holocène à l'Anthropocène: Intérêts Environnemental et Géoéconomique* [Contribution of Geochemical and Sedimentological Approaches to the Functioning of Endorheic and Exoreic Systems from the Holocene to the Anthropocene: Environmental and Geoeconomic Interests]. Institut Supérieur des Sciences et Techniques des Eaux de Gabès.

DeMenocal, P., Ortiz, J., Guilderson, T., & Sarnthein, M. (2000). Coherent high-and low-latitude climate variability during the Holocene warm period. *Science, 288*(5474), 2198–2202. doi:10.1126cience.288.5474.2198 PMID:10864866

Denizcilik Dergisi. (2021, April 15). *Danger of invasive species in the seas.* Denizcilik Dergisi [Maritime Magazine]. https://www.denizcilikdergisi.com/denizcilik-gundem-haberleri/denizlerde-istilaci-tur-tehlikesi/

DHA [Demirören News Agency]. (2018, September 28). Attention to 10 species dangerous for human health in the Mediterranean. *CNN Turkey.* https://www.cnnturk.com/turkiye/akdenizde-insan-sagligi-icin-tehlikeli-10-ture-dikkat

Dixon, M. J. R., Loh, J., Davidson, N. C., Beltrame, C., Freeman, R., & Walpole, M. (2016). Tracking global change in ecosystem area: The Wetland Extent Trends index. *Biological Conservation, 193*, 27–35. doi:10.1016/j.biocon.2015.10.023

Driouech, F. (2010). *Distribution des précipitations hivernales sur le Maroc dans le cadre d'un changement climatique: descente d'échelle et incertitudes.* [*Distribution of winter precipitation over Morocco in the context of climate change: downscaling and uncertainties.*] [Doctoral dissertation, Universite de Toulouse, France].

DWC. (2016). Vanadivi theth bim. *Science.gov.*

Eken, G. (1997). The breeding population of some species of waterbirds at Gediz Delta, Western Turkey. *Zoology in the Middle East, 14*(1), 53–68. doi:10.1080/09397140.1997.10637704

Erinç, S. (1955). *Gediz ve Büyük Menderes deltalarının morfolojisi* [Morphology of Gediz and Great Menderes deltas]. Türk Coğrafya Kur. Yay.

Ernoul, L., Sandoz, A., & Fellague, A. (2012). The evolution of two great Mediterranean Deltas: Remote sensing to visualize the evolution of habitats and land use in the Gediz and Rhone Deltas. *Ocean and Coastal Management, 69*, 111–117. doi:10.1016/j.ocecoaman.2012.07.026

Essefi, E. (2009). *Multidisciplinary study of Sidi El Hani Saline Environment: The Geological History and the Climatic Variability.* Academic Press.

Essefi, E. (2009). *Multidisciplinary study of Sidi El Hani Saline environment: the history and the climatic variability* [MS thesis]. Faculty of Sciences of Sfax, University of Sfax.

Essefi, E., Gharsalli, N., & Yaich, C. (2015c). Geophysical and geochemical study of the silico-evaporitic sedimentary filling of Boujmal wetland, eastern Tunisia: Inferring the climatic signal within groundwater noise. *J Basic Appl Res Int, 15*(2–1), 14–25.

Essefi, E., Gharsalli, N., Salhi, S., & Yaich, C. (2015b). Cyclicity of grain size parameters along a Core from the Gleysol of Sebkha Ennoual, southeastern Tunisia. *J Basic Appl Res Int, 15*(5–3), 124–134.

Essefi, E., Gharsalli, N., Kalabi, S., Ben Ameur, M., & Yaich, C. (2015d). Spectral analysis of a core from the Sebkha of Sidi Mansour, southern Tunisia: The Holocene cyclostratigraphy. *J Geophys Remote Sens*, *4*, 141. doi:10.4172/2169-0049.1000141

Essefi, E., Gharsalli, N., & Yaich, C. (2015a). Climatic significance aeolian sedimentary filling of sebkha Mhabeul, southeastern Tunisia: Contribution of the genetic approach. *J Basic Appl Res Int*, *15*(2–4), 140–152.

Essefi, E., Komatsu, G., Fairén, A. G., Chan, M. A., & Yaich, C. (2014). Groundwater influence on the aeolian sequence stratigraphy of the Mechertate–Chrita–Sidi El Hani system, Tunisian Sahel: Analogies to the wet-dry aeolian sequence stratigraphy at Meridiani Planum, Terby crater, and Gale crater, Mars. *Planetary and Space Science*, *95*, 56–78. doi:10.1016/j.pss.2013.05.010

Essefi, E., Komatsu, G., Fairén, A. G., Chan, M. A., & Yaich, C. (2014a). Models of formation and activity of spring mounds in the Mechertate-Chrita-Sidi El Hani system, eastern Tunisia: Implications for the habitability of Mars. *Life (Chicago, Ill.)*, *4*(3), 386–432. doi:10.3390/life4030386 PMID:25370379

Essefi, E., Mefteh, S., Medhioub, M., & Yaich, C. (2014). Magnetic study of the heated and unheated sedimentary fillings of Sebkha Mhabeul, Southeast Tunisia: A geophysical method for paleoclimatic investigation and tephrochronological dating. *International Journal of Geophysics*, *2014*, 2014. doi:10.1155/2014/908395

Essefi, E., Touir, J., Tagorti, M. A., & Yaich, C. (2013). Effect of the groundwater contribution, the climatic change, and the human-induced activities on the hydrological behavior of discharge playas: A case study Sidi El Hani discharge playa, Tunisian Sahel. *Arabian Journal of Geosciences*, *6*(10), 3997–4009. doi:10.100712517-012-0659-6

Essefi, E., Touir, J., Tagorti, M. A., & Yaich, C. (2013). Record of the climatic variability and the sedimentary dynamics during the last two millennia at sebkha Dkhila, eastern Tunisia. *International Scholarly Research Notices*.

Essefi, E., Touir, J., Tagorti, M. A., & Yaich, C. (2014b). Geodynamic framework of saline systems in eastern Tunisia: Saline depressions inherited from the Triassic intrusions and/or the Messinian Salinity Crisis. *International Scholarly Research Notices*.

European Union. (2019). *The EU Fish Market*. Publications Office of the European Union. https://www.eumofa.eu/documents/20178/314856/EN_The+EU+fish+market_2019.pdf/

FAO - Fisheries and Aquaculture Information and Statistics Branch. (2021), *Statistical Query Results, Capture Quantity*. FAO. http://www.fao.org/figis/servlet/SQServlet?file=/usr/local/tomcat/8.5.16/figis/webapps/figis/temp/hqp_4241499935846662994.xml&outtype=html

FAO (n.d.). *Fisheries statistics*. [Data set]. FAO. http://www.fao.org/fishery/facp/TUR/en#CountrySector-Statistics

FAO. (n.d.a). *Fishery and Aquaculture Country Profiles The Republic of Turkey*. FAO. https://www.fao.org/fishery/facp/TUR/en

FAO. (n.d.b). *GFCM AVL - Authorized Vessel List (>15m LOA)*. FAO. https://www.fao.org/gfcm/data/fleet/avl

FAO. (n.d.c). *General Fisheries Commission for the Mediterranean – GFCM*. FAO. https://www.fao.org/gfcm/data/fleet/blackseaturbot

FAO. (n.d.d). *Vessels authorized to fish for red shrimp and blue and red shrimp in the Levant Sea*. FAO. https://www.fao.org/gfcm/data/fleet/levantsea

Farid, H. U., Khan, Z. M., Shakoor, A., Mubeen, M., Ayub, H. U., Kanwar, R. M. A., & Bilal, M. (2022). Water Resources in Relation to Climate Change. In *Building Climate Resilience in Agriculture* (pp. 145–166). Springer. doi:10.1007/978-3-030-79408-8_10

Faust, D., Zielhofer, C., Escudero, R. B., & del Olmo, F. D. (2004). High-resolution fluvial record of late Holocene geomorphic change in northern Tunisia: Climatic or human impact? *Quaternary Science Reviews*, *23*(16-17), 1757–1775. doi:10.1016/j.quascirev.2004.02.007

Fernando, P. (2010). Disaster management measures of microfinance. *Sri Lanka Journal of Advanced Social Studies*, *2*, 30.

Finlayson, C. M., Davidson, N. C., Spiers, A. G., & Stevenson, N. J. (1999). Global wetland inventory–current status and future priorities. *Marine and Freshwater Research*, *50*(8), 717–727. doi:10.1071/MF99098

Finlayson, C., Milton, G., & Prentice, R. (2016). Wetland types and distribution. In C. Finlayson, G. Milton, R. Prentice, & N. Davidson (Eds.), *The wetland book*. Springer. From., doi:10.1007/978-94-007-6173-5_186-1

Finley, T. (2020). *Fault-hosted geothermal systems in southeastern British Columbia*. Academic Press.

Fletcher, W. J., & Zielhofer, C. (2013). Fragility of Western Mediterranean landscapes during Holocene rapid climate changes. *Catena*, *103*, 16–29. doi:10.1016/j.catena.2011.05.001

Freshwaterblog. (2021). The freshwater blog is the voice of aquatic life. *Fresh Water Blog*. https://freshwaterblog.net/2021/03/10/towards-a-freshwater-ethic-lessons-from-aldo-Leopold-for-contemporary-aquatic-conservation/

Galewski, T., & Devictor, V. (2016). When common birds became rare: Historical records shed light on long-term responses of bird communities to global change in the largest wetland of France. *PLoS One*, *11*(11), e0165542.GDNCNP (2020)

Galewski, T., Collen, B., McRae, L., Loh, J., Grillas, P., Gauthier-Clerc, M., & Devictor, V. (2011). Long-term trends in the abundance of Mediterranean wetland vertebrates: from global recovery to localized declines. *Biological Conservation*.

Galil, B. S. (2007). Loss or gain? Invasive aliens and biodiversity in the Mediterranean Sea. *Marine Pollution Bulletin*, *55*(7–9), 314–322. doi:10.1016/j.marpolbul.2006.11.008 PMID:17222869

Galil, B. S. (2008). Alien species in the Mediterranean Sea—which, when, where, why? In *Challenges to marine ecosystems* (pp. 105–116). Springer. doi:10.1007/978-1-4020-8808-7_10

Galil, B. S. (2009). Taking stock: Inventory of alien species in the Mediterranean Sea. *Biological Invasions*, *11*(2), 359–372. doi:10.100710530-008-9253-y

Gammoudi, A., Essefi, E., & Rigane, H. (2019). Variability of the grain-sized fractions during the last three millennia within the sedimentary fillings of the saline system of Mhabeul: The Holocene cyclostratigraphy. *Arabian Journal of Geosciences*, *12*(9), 1–11. doi:10.100712517-019-4472-3

Gardner, R. C., & Finlayson, M. (2018). Global wetland outlook: State of the world's wetlands and their services to people 2018. *Secretariat of the Ramsar Convention.*

Garnett, T., Appleby, M. C., Balmford, A., Bateman, I. J., Benton, T. G., Bloomer, P., Burlingame, B., Dawkins, M., Dolan, L., Fraser, D., Herrero, M., Hoffmann, I., Smith, P., Thornton, P. K., Toulmin, C., Vermeulen, S. J., & Godfray, H. C. J. (2013). Sustainable intensification in agriculture: Premises and policies. *Science*, *341*(6141), 33–34. doi:10.1126cience.1234485 PMID:23828927

Gediz Delta Management Plan. (2007). *Gediz Deltası Sulak Alan Yönetim Planı* [Gediz Delta Management Plan].

Geijzendorffer, I., Chazée, L., Gaget, E., Galewski, T., Guelmami, A., Perennou, C., & McInnes, R. (2018). Mediterranean wetlands outlook 2: Solutions for sustainable Mediterranean Wetlands. *Secretariat of the Ramsar Convention.*

Gil-Alana, L. A., & Yaya, O. S. (2014). The relationship between oil prices and the Nigerian stock market. An analysis based on fractional integration and cointegration. *Energy Economics*, *46*, 328–333. doi:10.1016/j.eneco.2014.10.001

Giorgi, F. (2006). Climate change hot-spots. *Geophysical Research Letters*, *33*(8), L08707. doi:10.1029/2006GL025734

Giraudeau, J., Cremer, M., Manthé, S., Labeyrie, L., & Bond, G. (2000). Coccolith evidence for instabilities in surface circulation south of Iceland during Holocene times. *Earth and Planetary Science Letters*, *179*(2), 257–268. doi:10.1016/S0012-821X(00)00113-8

Goddard, E. N., Trask, P. D., de Ford, R. K., Rove, O. N., Singewald, J. T., & Overbeck, R. M. (1951). *Rock Color Chart*. Geol. Soc.

Gokce, D. (2018). Wetland importance and management. In *Wetland management-Assessing risk and sustainable solutions* (pp. 1–9). Intech Open.

Goodwin, E. J. (2017). Convention on Wetlands of International Importance, Especially as Waterfowl Habitat 1971 (Ramsar). In Elgar Encyclopedia of Environmental Law (pp. 101-108). Edward Elgar Publishing Limited.

Gorelick, N., Hancher, M., Dixon, M., Ilyushchenko, S., Thau, D., & Moore, R. (2017). Google Earth Engine: Planetary-scale geospatial analysis for everyone. *Remote Sensing of Environment*, *202*, 18–27. doi:10.1016/j.rse.2017.06.031

Goren, M., Galil, B. S., Diamant, A., Stern, N., & Levitt-Barmats, Y. A. (2016). Invading up the food web? Invasive fish in the southeastern Mediterranean Sea. *Marine Biology*, *163*(8), 1–11. doi:10.100700227-016-2950-7

GoSL. (2016). *Final Metro Colombo Wetland Management Strategy*. Landscape and urban design office.

GoSL. (2018). *State of Sri Lankan Cities*. UN-Habitat.

Government of Canada. (n.d.). *A Canadian Action Plan to Address the Threat of Aquatic Invasive Species*. Government of Canada. https://www.dfo-mpo.gc.ca/species-especes/publications/ais-eae/plan/index-eng.html

Gray, D. M., Mckay, G. A., & Wigam, J. M. (1970). Energy, evaporation, and evapotranspiration. In D. M. Gray (Ed.), *Handbook on the Principles of Hydrology, Sect. Ill* (pp. 3.1–3.66). Water Information Center.

Grillas, P., Rhazi, L., Lefebvre, G., El Madihi, M., & Poulin, B. (2021). Foreseen impact of climate change on temporary ponds located along a latitudinal gradient in Morocco. *Inland Waters*, *11*(4), 492–507. doi:10.1080/20442041.2021.1962688

Gunasekara, T. (2010 May 12). The development war-a 'humanitarian operation' against Colombo's poor. *The Sunday Leader*.

Gunathilaka, M. D. K. L. (2020). Land use and land cover changes and Avifauna: An empirical analysis of loss of agricultural wetlands and its impact on avian species in suburban areas. *International Journal of Scientific and Research Publications*, *10*(5), 263–272. doi:10.29322/IJSRP.10.05.2020.p10132

Haber, I. (2020, August 18). *Invasive 2 Species Take Place Among Mediterranean Fish*. İklim Haber. https://ww.iklimhaber.org/istilaci-2-tur-akdeniz-baliklari-arasindaki-yerini-aldi/

Hajji, S., Ayed, B., Riahi, I., Allouche, N., Boughariou, E., & Bouri, S. (2018). Assessment and mapping groundwater quality using hybrid PCA-WQI model: Case of the Middle Miocene aquifer of Hajeb Layoun-Jelma basin (Central Tunisia). *Arabian Journal of Geosciences*, *11*(20), 1–21. doi:10.100712517-018-3924-5

Hajji, S., Karoui, S., Nasri, G., Allouche, N., & Bouri, S. (2021). EFA-CFA integrated approach for groundwater resources sustainability in agricultural areas under data scarcity challenge: Case study of the Souassi aquifer, Central-eastern Tunisia. *Environment, Development and Sustainability*, *23*(8), 1–20. doi:10.100710668-020-01155-5

Hajji, S., Nasri, G., Boughariou, E., Bahloul, M., Allouche, N., & Bouri, S. (2020). Towards understanding groundwater quality using hydrochemical and statistical approaches: Case of shallow aquifer of Mahdia–KsourEssaf (Sahel of Tunisia). *Environmental Science and Pollution Research International*, *27*(5), 5251–5265. doi:10.100711356-019-06982-2 PMID:31848945

Hamed, Y. (2011). The hydrogeochemical characterization of groundwater in Gafsa-Sidi Boubaker region (Southwestern Tunisia). *Arabian Journal of Geosciences*. Advance online publication. doi:10.100712517-011-0393-5

Hannam, J. A., & Dearing, J. A. (2008). Mapping soil magnetic properties in Bosnia and Herzegovina for landmine clearance operations. *Earth and Planetary Science Letters*, *274*(3-4), 285–294. doi:10.1016/j.epsl.2008.05.006

Hartmann, D. (2007). From reality to model: Operationalism and the value chain of particle-size analysis of natural sediments. *Sedimentary Geology*, *202*(3), 383–401. doi:10.1016/j.sedgeo.2007.03.013

Hay, K. L., Dearing, J. A., Baban, S. M. J., & Loveland, P. (1997). A preliminary attempt to identify atmospherically-derived pollution particles in English topsoils from magnetic susceptibility measurements. *Physics and Chemistry of the Earth*, *22*(1-2), 207–210. doi:10.1016/S0079-1946(97)00104-3

Heller, F., & Tung-sheng, L. (1986). Palaeoclimatic and sedimentary history from magnetic susceptibility of loess in China. *Geophysical Research Letters*, *13*(11), 1169–1172. doi:10.1029/GL013i011p01169

Henkanaththegedara, S. (2005). Ichthyofauna of its Bellanwila-Attidiya sanctuary and its environs in Colombo, Sri Lanka. *TriggerPaper, 32*(1).

Herath, S., & Rathnayake, U. (2005). Changing rainfall and its impact on landslides in Sri Lanka. *Journal of Mountain Science*, *2*(3), 2018–2224.

Hettiarachchi, P. (2018). Flood frequency analysis and selection of suitable design flood for Kelani flood bunds. *Research Gate.* https://www.researchgate.net/publication/327745195_Flood_Frequency_Analysis_for_the_Kelani_River_and_the_Design_Flood_for_Kelani_Flood_Bunds

Hettiarachchi, M., Athukoral, K., Wijekoon, S., & Alwis, A. (2014). Urban wetlands and disaster resilience of Colombo, Sri Lanka. *International Journal of Disaster Resilience in the Built Environment*, *5*(1), 79–89. doi:10.1108/IJDRBE-11-2011-0042

Hettiarachchi, M., Morrison, T., & McAlpine, C. (2017). Power, politics and policy in the appropriation of urban wetlands: The critical case of Sri Lanka. *The Journal of Peasant Studies*, 1–19.

Hettiarachchi, M., Morrison, T., Wickramasinghe, R., Mapa, A., & McApine, C. (2014). The eco-social transformation of urban wetlands: A case study of Colombo, Sri Lanka. *Landscape and Urban Planning*, *132*, 55–68. doi:10.1016/j.landurbplan.2014.08.006

Hewavitharana, B. (2004). Poverty alleviation. In S. Kelegama (Ed.), *Economic policy in Sri Lanka: Issues and debates* (pp. 467–497). Sage Publication.

Hollis, G.E. & Kallel, M.R. (1985). *Modelling natural and human induced hydrological changes on Sebkhet El Kelbia.* Academic Press.

Hsü, K. J. KJ, H., MB, C., & WBF, R. (1973). The origin of the Mediterranean evaporites. Academic Press.

Huang, C., Rao, Z., Li, Y., Yang, W., Liu, L., Zhang, X., & Wu, Y. (2020). Holocene summer temperature in arid central Asia linked to millennial-scale North Atlantic climate events and driven by centennial-scale solar activity. *Palaeogeography, Palaeoclimatology, Palaeoecology, 556*, 109880. doi:10.1016/j.palaeo.2020.109880

Hu, S., Niu, Z., Chen, Y., Li, L., & Zhang, H. (2017). Global wetlands: Potential distribution, wetland loss, and status. *The Science of the Total Environment, 586*, 319–327. doi:10.1016/j.scitotenv.2017.02.001 PMID:28190574

IPCC: Climate Change. (2014). *Synthesis Report. Contribution of working groups I. II and III to the fifth assessment report of the intergovernmental panel on climate change.* IPCC.

IUCN [International Union For Conservation Of Nature]. (n.d.). *Marine Menace, Alien invasive species in the marine environment.* IUCN. https://www.cbd.int/invasive/doc/marine-menace-iucn-en.pdf

IUCN. (2006). *CEA.* National Wetland Directory of Sri Lanka.

Jalali, B., Sicre, M. A., Bassetti, M. A., & Kallel, N. (2016). Holocene climate variability in the north-western Mediterranean Sea (gulf of lions). *Climate of the Past, 12*(1), 91–101. doi:10.5194/cp-12-91-2016

Jaouadi, S., Lebreton, V., Bout-Roumazeilles, V., Siani, G., Lakhdar, R., Boussoffara, R., Dezileau, L., Kallel, N., Mannai-Tayech, B., & Combourieu-Nebout, N. (2016). Environmental changes, climate and anthropogenic impact in south-east Tunisia during the last 8 kyr. *Climate of the Past, 12*(6), 1339–1359. doi:10.5194/cp-12-1339-2016

Jenkins, C. (2002). Automated digital mapping of geological color descriptions. *Geo-Marine Letters, 22*(4), 181–187. doi:10.100700367-002-0111-0

Jian, Z., Wang, P., Saito, Y., Wang, J., Pflaumann, U., Oba, T., & Cheng, X. (2000). Holocene variability of the Kuroshio current in the Okinawa Trough, northwestern Pacific Ocean. *Earth and Planetary Science Letters, 184*(1), 305–319. doi:10.1016/S0012-821X(00)00321-6

Jiménez-Espejo, F. J., García-Alix, A., Jiménez-Moreno, G., Rodrigo-Gámiz, M., Anderson, R. S., Rodríguez-Tovar, F. J., Martínez-Ruiz, F., Giralt, S., Delgado Huertas, A., & Pardo-Igúzquiza, E. (2014). Saharan aeolian input and effective humidity variations over western Europe during the Holocene from a high altitude record. *Chemical Geology, 374*, 1–12. doi:10.1016/j.chemgeo.2014.03.001

Kabii, T. (n.d.).An overview of African wetlands. *Aquadocs.* https://aquadocs.org/bitstream/handle/1834/457/Africa_Wetlands Retrieved Dec. 9. 2021.

Karunathilaka, K., Dabare, H., & Nandalal, K. (2017). Changes in rainfall in Sri Lanka during 1966-2015. *Journal of the Institution of Engineers, Sri Lanka*, *50*(2), 39–48. doi:10.4038/engineer.v50i2.7251

Katsanevakis, S., Coll, M., Piroddi, C., Steenbeek, J., Ben Rais Lasram, F., Zenetos, A., & Cardoso, A. C. (2014). Invading the Mediterranean Sea: Biodiversity patterns shaped by human activities. *Frontiers in Marine Science*, *1*, 32. doi:10.3389/fmars.2014.00032

Katsanevakis, S., Tempera, F., & Teixeira, H. (2016). Mapping the impact of alien species on marine ecosystems: The Mediterranean Sea case study. *Diversity & Distributions*, *22*(6), 694–707. doi:10.1111/ddi.12429

Kawamura, M., Kominami, K., & Takeuchi, J., & A., T.-e. (1996). A multicopy suppressor of nin1-1 of the yeast *Saccharomyces cerevisiae* is a counterpart of the Drosophila melanogaster diphenol oxidase A2 gene, Dox-A2. *Molecular & General Genetics*, *251*(0), 146–152. PMID:8668124

Koei, N. (2002). *The study on stormwater drainage plan for the Colombo metropolitan region in the democratic socialist republic of Sri Lanka.* Sri Lanka Land Reclamation and Development Corporation, JAICA.

Konikow, L. F. (2002, October). Ground-water depletion and overexploitation; a global problem. *Geological Society of America Abstracts with Programs*, *34*, 229.

Korunan Alanlar İstatistikleri (2019). *Protected Area Statistics.* [Data set]. Turkey Ministry of Agriculture and Forestry. https://www.tarimorman.gov.tr/DKMP/Menu/18/Korunan-Alan-Ista tistikleri#>.

Kotagama, S. W., & Bambaradeniya, C. B. (2006). An Overview of the Wetlands of Sri Lanka and their Conservation Significance. In National Wetland Directory of Sri Lanka (pp. 7-16). The Central Environmental Authority (CEA), The World Conservation Union (IUCN),the International Water Management Institute (IWMI).

Kourantidou, M., Cuthbert, R., Haubrock, P., Novoa, A., Taylor, N., Leroy, B., Capinha, C., Renault, D., Angulo, E., Diagne, C., & Courchamp, F. (2021). Economic costs of invasive alien species in the Mediterranean basin. *NeoBiota*, *67*, 427–458. doi:10.3897/neobiota.67.58926

Lakhdar, A., Rabhi, M., Ghnaya, T., Montemurro, F., Jedidi, N., & Abdelly, C. (2009). Effectiveness of compost use in salt-affected soil. *Journal of Hazardous Materials*, *171*(1-3), 29–37. doi:10.1016/j.jhazmat.2009.05.132 PMID:19576686

Li, N., Sharifi, A., Chambers, F. M., Ge, Y., Dubois, N., Gao, G., Li, D., Liu, L., Liu, H., Wang, J., Niu, H., Meng, M., Liu, Y., Zhang, G., & Jie, D. (2021). Linking Holocene East Asian monsoon variability to solar forcing and ENSO activity: Multi-proxy evidence from a peatland in Northeastern China. *The Holocene*, *31*(6), 966–982. doi:10.1177/0959683621994662

Lionello, P., Malanotte-Rizzoli, P., Boscolo, R., Alpert, P., Artale, V., Li, L., ... Xoplaki, E. (2006). The Mediterranean climate: an overview of the main characteristics and issues. *Developments in Earth and Environmental Sciences, 4*, 1-26.

Maher, B. A., & Thompson, R. (1995). Paleorainfall reconstructions from pedogenic magnetic susceptibility variations in the Chinese loess and paleosols. *Quaternary Research, 44*(3), 383–391. doi:10.1006/qres.1995.1083

Mamou, A., & Kassah, A. (2002). Eau et développement dans le sud tunisien [Water and development in southern Tunisia]. Cahier du CERES, Série Géographique no. 23.

Mannino, A. M., & Balistreri, P. (2021). Invasive alien species in Mediterranean Marine Protected Areas: The Egadi Islands (Italy) case study. *Biodiversity (Nepean), 22*(1-2), 13–23. doi:10.1080/14888386.2021.1875263

Marawila, T., & Tibbotuwawa, M. (2010). *To develop or to conserve: the case of the Diyawanna Oya wetlands in Sri Lanka*. South Asian Network for Development and Environmental Economics.

Marchini, A., Galil, B. S., & Occhipinti-Ambrogi, A. (2015). Recommendations on standardizing lists of marine alien species: Lessons from the Mediterranean Sea. *Marine Pollution Bulletin, 101*(1), 267–273. doi:10.1016/j.marpolbul.2015.09.054 PMID:26471066

Marignani, M., Lussu, M., Murru, V., Bacaro, G., & Cogoni, A. (2020). Effect of Invasive Alien Species on the Co-Occurrence Patterns of Bryophytes and Vascular Plant Species—The Case of a Mediterranean Disturbed Sandy Coast. *Diversity (Basel), 12*(4), 160. doi:10.3390/d12040160

Marquer, L., Pomel, S., Abichou, A., Schulz, E., Kaniewski, D., & Van Campo, E. (2008). Late Holocene high-resolution palaeoclimatic reconstruction inferred from Sebkha Mhabeul, southeast Tunisia. *Quaternary Research, 70*(2), 240–250. doi:10.1016/j.yqres.2008.06.002

Martinez-Ruiz, F., Kastner, M., Gallego-Torres, D., Rodrigo-Gámiz, M., Nieto-Moreno, V., & Ortega-Huertas, M. (2015). Paleoclimate and paleoceanography over the past 20,000 yr in the Mediterranean Sea Basins as indicated by sediment elemental proxies. *Quaternary Science Reviews, 107*, 25–46. doi:10.1016/j.quascirev.2014.09.018

Matthews, G. (2013). The Ramsar Convention on wetlands: its history and development (re- issue of 1993 edition ed.). Ramsar Convention Bureau.

Mayewski, P. A., Rohling, E. E., Stager, J. C., Karlén, W., Maasch, K. A., Meeker, L. D., ... Steig, E. J. (2004). Holocene climate variability. *Quaternary Research, 62*(3), 243–255. doi:10.1016/j.yqres.2004.07.001

McManus, J. F., Oppo, D. W., & Cullen, J. L. (1999). A 0.5-million-year record of millennial-scale climate variability in the North Atlantic. *Science, 283*(5404), 971-975.

MCUDP. (2016). *Metro Colombo Urban Development Project*. SLLRDC.

Mediterranean Wetland Observatory. (2014). *Land cover: Spatial dynamics in Metierranean coastal wetlands from 1975 to 2005.Thematic collection 2*. Tour du Valat.

Mefteh, S., Essefi, E., Yaich, C., Jamoussi, F., & Medhioub, M. (2015). Correlation between magnetic susceptibility and mineral species along NWA-1 well, southern Tunisia: An overlap of the depositional environment, the climate, and the diagenesis. *Journal of African Earth Sciences*, *103*, 89–101. doi:10.1016/j.jafrearsci.2014.12.007

Mitsch, W. J., & Gosselink, J. G. (2000). The value of wetlands: Importance of scale and landscape setting. *Ecological Economics*, *35*(1), 25–33. doi:10.1016/S0921-8009(00)00165-8

MoEF. (2020). A conceptual model to understand the drivers of change in tropical wetlands: A comparative assessment in India and Brazil. *Biota neotropical*. doi:10.1590/1676-BN-2019-09-13

Mokadem, N. (2012). Hydrodynamic and mineralization of deep groundwater in El Guettar–Gafsa basin ''SW Tunisia'' [Master]. Faculty of Sciences of Bizerte.

Mokadem, N., Hamed, Y., Hfaid, M., & Dhia, H. B. (2015). Hydrogeochemical and isotope evidence of groundwater evolution in El Guettar Oasis area, Southwest Tunisia. *Carbonates and Evaporites*, *30*(4), 417–437. doi:10.100713146-015-0235-8

Mokadem, N., Redhaounia, B., Besser, H., Ayadi, Y., Khelifi, F., Hamad, A., Hamed, Y., & Bouri, S. (2018). Impact of climate change on groundwater and the extinction of ancient "Foggara" and springs systems in arid lands in North Africa: A case study in Gafsa basin (Central of Tunisia). *Euro-Mediterranean Journal for Environmental Integration*, *3*(1), 1–14. doi:10.100741207-018-0070-0

Mombauer, D. (2019). Colombo's urban wetlands: on becoming and staying a Ramsar wetland city. *Slycan Trust.*

Moy, C. M., Seltzer, G. O., Rodbell, D. T., & Anderson, D. M. (2002). Variability of El Niño/ Southern Oscillation activity at millennial timescales during the Holocene epoch. *Nature*, *420*(6912), 162–165. doi:10.1038/nature01194 PMID:12432388

Munter, J. A., & Anderson, M. P. (1981). The use of ground-water flow models for estimating lake seepage rates. *Ground Water*, *19*(6), 608–616. doi:10.1111/j.1745-6584.1981.tb03515.x

Nielsen, S. H., Koç, N., & Crosta, X. (2004). Holocene climate in the Atlantic sector of the Southern Ocean: Controlled by insolation or oceanic circulation? *Geology*, *32*(4), 317–320. doi:10.1130/G20334.1

Nilguuml; l, K., Alpay, T., & Emel, B. (2009). Wetland management in Turkey: Problems, achievements and perspectives. *African Journal of Agricultural Research, 4*(11), 1106-1119.

NOAA Fisheries [National Oceanic and Atmospheric Administration]. (n.d.). *Invasive and Exotic Marine Species.* NOAA. https://www.fisheries.noaa.gov/insight/invasive-and-exotic-marine-species

Obrochta, S. P., Miyahara, H., Yokoyama, Y., & Crowley, T. J. (2012). A re-examination of evidence for the North Atlantic "1500-year cycle" at Site 609. *Quaternary Science Reviews*, *55*, 23–33. doi:10.1016/j.quascirev.2012.08.008

Occhipinti-Ambrogi, A. (2007). Global change and marine communities: Alien species and climate change. *Marine Pollution Bulletin, 55*(7-9), 342–352. doi:10.1016/j.marpolbul.2006.11.014 PMID:17239404

Occhipinti-Ambrogi, A., Marchini, A., Cantone, G., Castelli, A., Chimenz, C., Cormaci, M., Froglia, C., Furnari, G., Gambi, M. C., Giaccone, G., Giangrande, A., Gravili, C., Mastrototaro, F., Mazziotti, C., Orsi-Relini, L., & Piraino, S. (2011). Alien species along the Italian coasts: An overview. *Biological Invasions, 13*(1), 215–237. doi:10.100710530-010-9803-y

Odum, E. (2002). Tidal marshes as out welling/pulsing systems. In M. Weinstein & D. Kreeger (Eds.), *Concepts and Controversies in Tidal Marsh Ecology* (pp. 3–7). Kluwer Academic. doi:10.1007/0-306-47534-0_1

Ojala, A. E. K., Launonen, I., Holmström, L., & Tiljander, M. (2015). Effects of solar forcing and North Atlantic oscillation on the climate of continental Scandinavia during the Holocene. *Quaternary Science Reviews, 112*, 153–171. doi:10.1016/j.quascirev.2015.01.021

Özacar, İ. U. (2019, September 26). *Fish stocks are declining in the seas.* TRT Haber. https://www.trthaber.com/haber/turkiye/denizlerde-balik-stoku-azaliyor-432836.html

Özacar, İ. U. (2020, September 2). *Climate change is also increasing invasive species in the seas.* TRT Haber. https://www.trthaber.com/haber/turkiye/iklim-degisikligi-denizlerdeki-istilaci-turleri-de-artiriyor-513312.html

Panditaratna, B. (1965). The trend of industrialization in Colombo city: the Capital of Ceylon. *Pakistan Geographical Review, 20*(2).

Patel, N. N., Angiuli, E., Gamba, P., Gaughan, A., Lisini, G., Stevens, F. R., Tatem, A. J., & Trianni, G. (2015). Multitemporal settlement and population mapping from Landsat using Google Earth Engine. *International Journal of Applied Earth Observation and Geoinformation, 35*, 199–208. doi:10.1016/j.jag.2014.09.005

Pathirana, A., Hailu, B., William, V., Chris, Z., & Allan, B. (2014). Impact of urban growth- driven land-use change on microclimate and extreme precipitation - A sensitivity study. *Atmospheric Research, 20*(2), 59–72. doi:10.1016/j.atmosres.2013.10.005

Pekel, J. F., Cottam, A., Gorelick, N., & Belward, A. S. (2016). High-resolution mapping of global surface water and its long-term changes. *Nature, 540*(7633), 418–422. doi:10.1038/nature20584 PMID:27926733

Perennou, C., Guelmami, A., Paganini, M., Philipson, P., Poulin, B., Strauch, A., Tottrup, C., Truckenbrodt, J., & Geijzendorffer, I. R. (2018). Mapping Mediterranean wetlands with remote sensing: A good-looking map is not always a good map. *Advances in Ecological Research, 58*, 243–277. doi:10.1016/bs.aecr.2017.12.002

PMV. (2008). *Plan Maroc Vert* [*Green Morocco map*]. MAPMDREF.

Postel, S. (2019). Recent Loss of Freshwater Wetlands Worldwide Valued at 27 Trillion a Year. *National Geographic*. https://blog.nationalgeographic.org/2014/06/24/recent-loss-of-freshwater- wetlands-worldwide-valued-at-2-7-trillion-per-year

Poudel, D. D., Lee, T., Srinivasan, R., Abbaspour, K., & Jeong, C. Y. (2013). Assessment of seasonal and spatial variation of surface water quality, identification of factors associated with water quality variability, and the modeling of critical nonpoint source pollution areas in an agricultural watershed. *Journal of Soil and Water Conservation*, *68*(3), 155–171. doi:10.2489/jswc.68.3.155

Ramsar Convention on Wetlands. (2018). *Global Wetland Outlook: State of the World's Wetlands and their Services to People. Ramsar Convention Secretariat, Gland, Switzerland Ramsar Convention Secretariat. (2016). The fourth strategic plan 2016-2024. Ramsar handbooks for wise use of wetlands*. Ramsar Convention Secretariat.

Ranasinghe, P. (2018). *Ramsar accreditation of Colombo Wetlands: Rebranding Colombo as a wetland city*. Daily FT.

Ranasinghe, P., Piyadasa, R., & Mataraarachchi, S. (2019). A study to identify the soil characteristics of wetlands in Colombo, Sri Lanka. Dubai: *13th Wetland Conference*.

Reddy, K. R., & DeLaune, R. D. (2008). *Biogeochemistry of Wetland*. Tyler and Frances group.

Retallack, G. J., Sheldon, N. D., Cogoini, M., & Elmore, R. D. (2003). Magnetic susceptibility of early Paleozoic and Precambrian paleosols. *Palaeogeography, Palaeoclimatology, Palaeoecology*, *198*(3-4), 373–380. doi:10.1016/S0031-0182(03)00479-6

Rhinaldo-Lee, M. B., & Anderson, M. P. (1980). High water levels in ground-water dominated lakes- a case study from north-western Wisconsin. *Ground Water*, *8*, 334–339.

Risebrobakken, B., Jansen, E., Andersson, C., Mjelde, E., & Hevrøy, K. (2003). A high-resolution study of Holocene paleoclimatic and paleoceanographic changes in the Nordic Seas. *Paleoceanography*, *18*(1), n/a. doi:10.1029/2002PA000764

Rodbell, D. T., Seltzer, G. O., Anderson, D. M., Abbott, M. B., Enfield, D. B., & Newman, J. H. (1999). A 15,000-year record of El Niño-driven alluviation in southwestern Ecuador. *Science*, *283*(5401), 516–520. doi:10.1126cience.283.5401.516 PMID:9915694

Rosen, M. R. (1994). *The importance of groundwater in playas: A review of playa classifications and the sedimentology and hydrology of playas*. Academic Press.

Samarasinhe, Y., & Dayawansa, N. (2013). Remote sensing and GIS-based study in an assessment of the degradation risk of the Kolonnawa marsh. *Journal of the National Science Foundation of Sri Lanka*, *41*(4), 327–335. doi:10.4038/jnsfsr.v41i4.6256

Sarı, İ. (2021, September 13). *Invasive species threaten Marmara and Black Sea as well! What awaits us if native fish species decrease?* Hurriyet. https://www.hurriyet.com.tr/gundem/istilaci-turler-marmara-ve-karadenizi-de-tehdit-ediyor-yerli-balik-turleri-azalirsa-bizi-neler-bekliyor-41885358

Sato, T., Takahashi, H. A., Kawabata, K., Takahashi, M., Inamura, A., & Handa, H. (2020, January 1). Changes in the nitrate concentration of spring water after the 2016 Kumamoto earthquake. *Journal of Hydrology (Amsterdam)*, *580*, 124310. doi:10.1016/j.jhydrol.2019.124310

Sayad, A., & Chakiri, S. (2010). Impact de l'évolution du climat sur le niveau de Dayet Aoua dans le Moyen Atlas marocain. [Impact of climate change on the level of Dayet Aoua in the Moroccan Middle Atlas.] *Science et changements planétaires/Sécheresse, 21*(4), 245-251.

Scaramelli, C. (2020). The Lost Wetlands of Turkey. *Middle East Report (New York, N.Y.)*, *296*, 14.

Scholz, M., & Lee, B. H. (2005). Constructed wetlands: A review. *The International Journal of Environmental Studies*, *62*(4), 421–447. doi:10.1080/00207230500119783

Scott, D. (1989). *A Directory of Asian Wetlands.* IUCN.

Sebbar, A. (2013). Etude de la variabilité et de l'évolution de la pluviométrie au Maroc (1935-2005): Réactualisation de la carte des précipitations. [Study of the variability and evolution of rainfall in Morocco (1935-2005): Updating of the rainfall map.] [Doctoral Dissertation, University of Hassan II Mohammedia, Turkey].

Şekercioğlu, Ç. H., Anderson, S., Akçay, E., Bilgin, R., Can, Ö. E., Semiz, G., & Dalfes, H. N. (2011). Turkey's globally important biodiversity in crisis. *Biological Conservation*, *144*(12), 2752–2769. doi:10.1016/j.biocon.2011.06.025

Servello, G., Andaloro, F., Azzurro, E., Castriota, L., Catra, M., Chiarore, A., Crocetta, F., D'Alessandro, M., Denitto, F., Froglia, C., Gravili, C., Langer, M. R., Lo Brutto, S., Mastrototaro, F., Petrocelli, A., Pipitone, C., Piraino, S., Relini, G., Serio, D., & Zenetos, A. (2019). Marine alien species in Italy: A contribution to the implementation of descriptor D2 of the marine strategy framework directive. *Mediterranean Marine Science*, *20*(1), 1–48. doi:10.12681/mms.18711

Sghari, A., & Mercier, J. L. (2008). Le Messinien de la formation "Segui" du Sahel tunisien: un équivalent des "Gilbert deltas" [The Messinian of the "Segui" formation of the Tunisian Sahel: an equivalent of the "Gilbert deltas"]. Séance spécialisée de la Société Géologique de France, 70.

Shine, C., & Klemm, De. C. (1999). Fisheries conservation and management and the conservation of marine biological diversity in developments in international fisheries law. Kluwer.

Sıkı, M. (2020). *Gediz deltasi'ndaki İzmir kuş cennetini yaşatma mücadelesi (1982-2018)* [The struggle to keep the Izmir bird paradise alive in the Gediz delta (1982-2018).]. Kanguru Yayınları.

Silva, G., Weerakoon, S., & Herath, S. (2016). Event-Based Flood Inundation Mapping Under the Impact of ClimateChange: A Case Study in Lower Kelani River Basin, Sri Lanka. *Hydrology: Current Research*, *7*(1). doi:10.4172/2157-7587.1000228

Smith, R. C., & Baker, K. S. (1981). Optical properties of the clearest natural waters (200–800 nm). *Applied Optics*, *20*(2), 177–184. doi:10.1364/AO.20.000177 PMID:20309088

Smith, R. L. (1980). *Ecology and Field Biology* (3rd ed.). Harper and Row.

Solanki, S. K., Usoskin, I. G., Kromer, B., Schüssler, M., & Beer, J. (2004). Unusual activity of the Sun during recent decades compared to the previous 11,000 years. *Nature*, *431*(7012), 1084–1087. doi:10.1038/nature02995 PMID:15510145

Sorto-Ortiz, L. (2015). The regulation of ecological communities through feedback loops: A. *Research in Zoology*, *5*, 1–15.

Sri Lanka Land Reclamation and Development Corporation. (2015). *Technical report 2 Ecological status.* [Data set]. Sri Lanka Land Reclamation and Development Corporation.

Stebich, M., Rehfeld, K., Schluetz, F., Tarasov, P. E., Liu, J., & Mingram, J. (2015). Holocene vegetation and climate dynamics of NE China based on the pollen record from Sihailongwan Maar Lake. *Quaternary Science Reviews*, *124*, 275–289. doi:10.1016/j.quascirev.2015.07.021

Stein, A., Gerstner, K., & Kreft, H. (2014). Environmental heterogeneity as a universal driver of species richness across taxa, biomes and spatial scales. *Ecology Letters*, *17*(7), 866–880. doi:10.1111/ele.12277 PMID:24751205

Steven, J., Abigail, J., John, J., Julian, D., Andrea, J., & Steve, J. (2021). Stewardship and management of freshwater ecosystems: From Leopold's land ethic to a freshwater ethic. *Aquatic Conservation*, 1–13.

Subasinghe, S., Estoque, R., & Murayama, Y. (2016). Spatiotemporal analysis of urban growth using GIS and Remote sensing: A case study of the Colombo metropolitan area, Sri Lanka. *International Journal of Geo-Information*, *5*(197), 197. doi:10.3390/ijgi5110197

Sulthana, P., & Thompson, P. (2010). *Natural resource conflicts and community organizations in Bangladesh.* Department of Fisheries.

Sushmita, S. (2012). Concrete invasion: Colombo's marshes cannot avert floods the way they used to. Construction is to be blamed. *Down to Earth.* https://www.downtoearth.org.in/coverage/urbanisation/concrete-invasion-35850

Swindles, G. T., Morris, P. J., Baird, A. J., Blaauw, M., & Plunkett, G. (2012). Ecohydrological feedbacks confound peat-based climate reconstructions. *Geophysical Research Letters*, *39*(11), n/a. doi:10.1029/2012GL051500

Tagorti, M. A. (1995). Rapport d'étude bibliographique sur le secteur de la sebkha de Sidi El Hani [Bibliographic study report on the sebkha sector of Sidi El Hani]. Academic Press.

Tagorti, M. A., Essefi, E., Touir, J., Guellala, R., & Yaich, C. (2013). Geochemical controls of groundwaters upwelling in saline environments: Case study the discharge playa of Sidi El Hani (Sahel, Tunisia). *Journal of African Earth Sciences*, *86*, 1–9. doi:10.1016/j.jafrearsci.2013.05.004

TekadeRupali, P. (2021). The Global Climate Changes and Its Impact on Aquatic Environment. International Journal for Innovative Research in Multidisciplinary Field, 282.

The Fish Site. (2021, April 28). *Turkey breaks $1 billion aquaculture export milestone*. The Fish Site. https://thefishsite.com/articles/turkey-breaks-1-billion-aquaculture-export-milestone

Torrent, J., & Barrón, V. (2002). Diffuse reflectance spectroscopy of iron oxides. Encyclopedia of Surface and Colloid Science, 1, 1438-1446.

TUIK TSI. (2020). Address Based Population Registration System Results, 2019. [Data set]. TurkStat.

TURKSTAT. (2021a). *Fishery statistics*. TurkStat. https://data.tuik.gov.tr/Kategori/GetKategori?p=tarim-111&dil=1

TURKSTAT. (2021b). *Quantity of caught sea fish*. TurkStat. https://data.tuik.gov.tr/Kategori/GetKategori?p=tarim-111&dil=1

TURKSTAT. (2021c). *Quantity of caught other sea fish (crustaceas, molluscs)*. TurkStat. https://data.tuik.gov.tr/Kategori/GetKategori?p=tarim-111&dil=1

TURKSTAT. (2021d). *Landing of sea fish for intended use, 2020*. TurkStat. https://data.tuik.gov.tr/Kategori/GetKategori?p=tarim-111&dil=1

TURKSTAT. (2021e). *Landing of other sea products for intended use, 2020*. TurkStat. https://data.tuik.gov.tr/Kategori/GetKategori?p=tarim-111&dil=1

UNDP. (1978). *Colombo master plan project*. Ministry of Local Government & Housing & Construction, Department of Town & Country Planning.

UNFCC. (2018, October 1). Wetlands disappear three times faster than forests. *United Nations Framework Convention on Climate Change*. https://unfccc.int/news/wetlands-disappearing-three-times-faster-than-forests

UoM. (2011). Environmental screening report for establishment of Beddagana and Kotte Ramparts Parks. *Sri Lanka - Metro Colombo Urban Development Project: environmental assessment*, 8.

Wanner, H., Beer, J., Bütikofer, J., Crowley, T. J., Cubasch, U., Flückiger, J., Goosse, H., Grosjean, M., Joos, F., Kaplan, J. O., Küttel, M., Müller, S. A., Prentice, I. C., Solomina, O., Stocker, T. F., Tarasov, P., Wagner, M., & Widmann, M. (2008). Mid-to Late Holocene climate change: An overview. *Quaternary Science Reviews*, *27*(19-20), 1791–1828. doi:10.1016/j.quascirev.2008.06.013

Warken, S. F., Schorndorf, N., Stinnesbeck, W., Hennhoefer, D., Stinnesbeck, S. R., Förstel, J., Steidle, S. D., Avilés Olguin, J., & Frank, N. (2021). Solar forcing of early Holocene droughts on the Yucatán peninsula. *Scientific Reports, 11*(1), 1–12. doi:10.103841598-021-93417-z PMID:34230539

Weber, N., Lazar, B., Gavrieli, I., Yechieli, Y., & Stein, M. (2021). Gypsum deltas at the Holocene Dead Sea linked to grand solar minima. *Geophysical Research Letters, 48*(6), e2020GL091034.

Weise, K., Höfer, R., Franke, J., Guelmami, A., Simonson, W., Muro, J., O'Connor, B., Strauch, A., Flink, S., Eberle, J., Mino, E., Thulin, S., Philipson, P., van Valkengoed, E., Truckenbrodt, J., Zander, F., Sánchez, A., Schröder, C., Thonfeld, F., & Hilarides, L. (2020). Wetland extent tools for SDG 6.6. 1 reporting from the Satellite-based Wetland Observation Service (SWOS). *Remote Sensing of Environment, 247*, 111892. doi:10.1016/j.rse.2020.111892

Wetzel, R. (2001). *Limnology: Lake and river ecosystems* (3rd ed.). Academic Press.

Winter, T. C., & Pfannkuch, H. O. (1984). Effect of anisotropy and groundwater system geometry on seepage through lakebeds: 12. Numerical simulation analysis. Journal of Hydrology (Amsterdam), 75(1-4), 239–253. doi:10.1016/0022-1694(84)90052-0 doi:10.1016/0022-1694(84)90052-0

Winter, T. U. (1986). Effect of ground-water recharge on the configuration of the water table beneath sand dunes and on seepage in lakes in the Sandhills of Nebraska. U.S.A. J. Hydrol., 86(3-4), 221–237. doi:10.1016/0022-1694(86)90166-6 doi:10.1016/0022-1694(86)90166-6

Wolanski, E., & Hopper, C. (2022). Dams and climate change accelerate channel avulsion and coastal erosion and threaten Ramsar-listed wetlands in the largest Great Barrier Reef watershed. Ecohydrology & Hydrobiology, 22(2), 197–212. doi:10.1016/j.ecohyd.2022.01.001 doi:10.1016/j.ecohyd.2022.01.001

WWF. (2018). *Puffer fish*. World Wildlife Foundation. https://www.wwf.org.tr/basin_bultenleri/guncel_haberler/?742
0/balonbaliklari

Xu, D., Lu, H., Chu, G., Liu, L., Shen, C., Li, F., Wang, C., & Wu, N. (2019). Synchronous 500-year oscillations of monsoon climate and human activity in Northeast Asia. *Nature Communications, 10*(1), 1–10. doi:10.103841467-019-12138-0 PMID:31511523

Xu, D., Lu, H., Chu, G., Shen, C., Li, F., Wu, J., Wang, L., Li, H., Yu, Y., Jin, Y., & Wu, N. (2020). Asynchronous 500-year summer monsoon rainfall cycles between Northeast and Central China during the Holocene. *Global and Planetary Change, 195*, 103324. doi:10.1016/j.gloplacha.2020.103324

Xu, D., Lu, H., Wu, N., Liu, Z., Li, T., Shen, C., & Wang, L. (2013). Asynchronous marine-terrestrial signals of the last deglacial warming in East Asia associated with low-and high-latitude climate changes. *Proceedings of the National Academy of Sciences of the United States of America, 110*(24), 9657–9662. doi:10.1073/pnas.1300025110 PMID:23720306

Yechieli, Y., & Wood, W. W. (2002). Hydrogeologic processes in saline systems: Playas, sabkhas, and saline lakes. Earth-Science Reviews, 58(3-4), 343–365. doi:10.1016/S0012-8252(02)00067-3 doi:10.1016/S0012-8252(02)00067-3

Yildirim, S., & Kaplan, M. (2022). The Threat of Invasive Alien Marine Species to the Blue Economy: The Mediterranean Case. In Raimi, L., & Kah, J. M. (Eds.). Implications for Entrepreneurship and Enterprise Development in the Blue Economy. IGI Global. doi:10.4018/978-1-6684-3393-5.ch004

Yıldırım, S., Yıldırım, D. Ç., & Gedikli, A. (2018). Sustainable Consumption Trends in the World in the Context of Green Economy and Sustainability. In I. Management Association (Ed.), Sustainable Development: Concepts, Methodologies, Tools, and Applications (pp. 1605-1624). IGI Global. doi:10.4018/978-1-5225-3817-2.ch071

Yildirim, S., Demirtas, I., & Yildirim, D. C. (2022). A Review of Alternative Economic Approaches to Achieve Sustainable Development: The Rising Digitalization and Degrowth Post COVID-19. In R. Castanho (Ed.), *Handbook of Research on Sustainable Development Goals, Climate Change, and Digitalization* (pp. 288–307). IGI Global. doi:10.4018/978-1-7998-8482-8.ch018

Yildirim, S., & Kaplan, M. (2020). Seafood Security and Sustainability Through Sustainable Development: A Review of Turkish Seafood Market. In A. Obayelu & O. Obayelu (Eds.), *Developing Sustainable Food Systems, Policies, and Securities* (pp. 16–36). IGI Global., doi:10.4018/978-1-7998-2599-9.ch002

Yıldırım, S., & Yıldırım, D. C. (2021). Achieving Seafood Security in the Mediterranean Region: A Case of Turkey. In R. Castanho & J. Martín Gallardo (Eds.), *Management and Conservation of Mediterranean Environments* (pp. 175–195). IGI Global., doi:10.4018/978-1-7998-7391-4.ch011

Zedler, J. B. B., & Kercher, S. (2005). Wetland Resources: Status, Trends. *Ecosystem Services, and Restorability. Annual Review of Environment and Resources*, *30*(1), 39–74. doi:10.1146/annurev.energy.30.050504.144248

Zenetos, A., Gofas, S., Morri, C., Rosso, A., Violanti, D., Raso, J. G., & Verlaque, M. (2012). Alien species in the Mediterranean Sea by 2012. A contribution to the application of European Union's Marine Strategy Framework Directive (MSFD). Part 2. Introduction trends and pathways. *Mediterranean Marine Science*, *13*(2), 328–352. doi:10.12681/mms.327

Zon, V. (2003). Wetlands conservation and management in Sri Lanka: A status paper. *National symposium on wetland conservation and management* (pp. 2-18). IUCN.

Related References

To continue our tradition of advancing academic research, we have compiled a list of recommended IGI Global readings. These references will provide additional information and guidance to further enrich your knowledge and assist you with your own research and future publications.

Ajiboye, O. E., & Yusuff, O. S. (2017). Foreign Land Acquisition: Food Security and Food Chains – The Nigerian Experience. In I. Management Association (Ed.), Natural Resources Management: Concepts, Methodologies, Tools, and Applications (pp. 1524-1545). Hershey, PA: IGI Global. https://doi.org/ doi:10.4018/978-1-5225-0803-8.ch072

Alapiki, H. E., & Amadi, L. A. (2018). Sustainable Food Consumption in the Neoliberal Order: Challenges and Policy Implications. In A. Obayelu (Ed.), *Food Systems Sustainability and Environmental Policies in Modern Economies* (pp. 90–123). Hershey, PA: IGI Global. doi:10.4018/978-1-5225-3631-4.ch005

Altaş, A. (2018). Geographical Information System Applications Utilized in Museums in Turkey Within the Scope of the Cultural Heritage Tourism: A Case Study of Mobile Application of Müze Asist. In S. Chaudhuri & N. Ray (Eds.), *GIS Applications in the Tourism and Hospitality Industry* (pp. 42–60). Hershey, PA: IGI Global. doi:10.4018/978-1-5225-5088-4.ch002

Andreea, I. R. (2018). Beyond Macroeconomics of Food and Nutrition Security. *International Journal of Sustainable Economies Management*, *7*(1), 13–22. doi:10.4018/IJSEM.2018010102

Anwar, J. (2017). Reproductive and Mental Health during Natural Disaster: Implications and Issues for Women in Developing Nations – A Case Example. In I. Management Association (Ed.), Gaming and Technology Addiction: Breakthroughs in Research and Practice (pp. 446-472). Hershey, PA: IGI Global. https://doi.org/ doi:10.4018/978-1-5225-0778-9.ch021

Awadh, H., Aksissou, M., Benhardouze, W., Darasi, F., & Snaiki, J. (2018). Socioeconomic Status of Artisanal Fishers in the West Part of Moroccan Mediterranean. *International Journal of Social Ecology and Sustainable Development*, *9*(1), 40–52. doi:10.4018/IJSESD.2018010104

Aye, G. C., & Haruna, R. F. (2018). Effect of Climate Change on Crop Productivity and Prices in Benue State, Nigeria: Implications for Food Security. In V. Erokhin (Ed.), *Establishing Food Security and Alternatives to International Trade in Emerging Economies* (pp. 244–268). Hershey, PA: IGI Global. doi:10.4018/978-1-5225-2733-6.ch012

Azizan, S. A., & Suki, N. M. (2017). Consumers' Intentions to Purchase Organic Food Products. In T. Esakki (Ed.), *Green Marketing and Environmental Responsibility in Modern Corporations* (pp. 68–82). Hershey, PA: IGI Global. doi:10.4018/978-1-5225-2331-4.ch005

Barakabitze, A. A., Fue, K. G., Kitindi, E. J., & Sanga, C. A. (2017). Developing a Framework for Next Generation Integrated Agro Food-Advisory Systems in Developing Countries. In I. Management Association (Ed.), Agri-Food Supply Chain Management: Breakthroughs in Research and Practice (pp. 47-67). Hershey, PA: IGI Global. doi:10.4018/978-1-5225-1629-3.ch004

Beachcroft-Shaw, H., & Ellis, D. (2018). Using Successful Cases to Promote Environmental Sustainability: A Social Marketing Approach. In I. Management Association (Ed.), Sustainable Development: Concepts, Methodologies, Tools, and Applications (pp. 936-953). Hershey, PA: IGI Global. doi:10.4018/978-1-5225-3817-2.ch042

Behnassi, M., Kahime, K., Boussaa, S., Boumezzough, A., & Messouli, M. (2017). Infectious Diseases and Climate Vulnerability in Morocco: Governance and Adaptation Options. In I. Management Association (Ed.), Public Health and Welfare: Concepts, Methodologies, Tools, and Applications (pp. 91-109). Hershey, PA: IGI Global. https://doi.org/ doi:10.4018/978-1-5225-1674-3.ch005

Bekele, F., & Bekele, I. (2017). Social and Environmental Impacts on Agricultural Development. In W. Ganpat, R. Dyer, & W. Isaac (Eds.), *Agricultural Development and Food Security in Developing Nations* (pp. 21–56). Hershey, PA: IGI Global. doi:10.4018/978-1-5225-0942-4.ch002

Benaouda, A., & García-Peñalvo, F. J. (2018). Towards an Intelligent System for the Territorial Planning: Agricultural Case. In F. García-Peñalvo (Ed.), *Global Implications of Emerging Technology Trends* (pp. 158–178). Hershey, PA: IGI Global. doi:10.4018/978-1-5225-4944-4.ch010

Bhaskar, A., Rao, G. B., & Vencatesan, J. (2017). Characterization and Management Concerns of Water Resources around Pallikaranai Marsh, South Chennai. In P. Rao & Y. Patil (Eds.), *Reconsidering the Impact of Climate Change on Global Water Supply, Use, and Management* (pp. 102–121). Hershey, PA: IGI Global. doi:10.4018/978-1-5225-1046-8.ch007

Bhyan, P., Shrivastava, B., & Kumar, N. (2022). Requisite Sustainable Development Contemplating Buildings: Economic and Environmental Sustainability. In A. Hussain, K. Tiwari, & A. Gupta (Eds.), *Addressing Environmental Challenges Through Spatial Planning* (pp. 269–288). IGI Global. https://doi.org/10.4018/978-1-7998-8331-9.ch014

Bogataj, D., & Drobne, D. (2017). Control of Perishable Goods in Cold Logistic Chains by Bionanosensors. In S. Joo (Ed.), *Applying Nanotechnology for Environmental Sustainability* (pp. 376–402). Hershey, PA: IGI Global. doi:10.4018/978-1-5225-0585-3.ch016

Bogataj, D., & Drobne, D. (2017). Control of Perishable Goods in Cold Logistic Chains by Bionanosensors. In I. Management Association (Ed.), Materials Science and Engineering: Concepts, Methodologies, Tools, and Applications (pp. 471-497). Hershey, PA: IGI Global. https://doi.org/ doi:10.4018/978-1-5225-1798-6.ch019

Bogueva, D., & Marinova, D. (2018). What Is More Important: Perception of Masculinity or Personal Health and the Environment? In D. Bogueva, D. Marinova, & T. Raphaely (Eds.), *Handbook of Research on Social Marketing and Its Influence on Animal Origin Food Product Consumption* (pp. 148–162). Hershey, PA: IGI Global. doi:10.4018/978-1-5225-4757-0.ch010

Bouzid, M. (2017). Waterborne Diseases and Climate Change: Impact and Implications. In M. Bouzid (Ed.), *Examining the Role of Environmental Change on Emerging Infectious Diseases and Pandemics* (pp. 89–108). Hershey, PA: IGI Global. doi:10.4018/978-1-5225-0553-2.ch004

Bowles, D. C. (2017). Climate Change-Associated Conflict and Infectious Disease. In M. Bouzid (Ed.), *Examining the Role of Environmental Change on Emerging Infectious Diseases and Pandemics* (pp. 68–88). Hershey, PA: IGI Global. doi:10.4018/978-1-5225-0553-2.ch003

Buck, J. J., & Lowry, R. K. (2017). Oceanographic Data Management: Quills and Free Text to the Digital Age and "Big Data". In P. Diviacco, A. Leadbetter, & H. Glaves (Eds.), *Oceanographic and Marine Cross-Domain Data Management for Sustainable Development* (pp. 1–22). Hershey, PA: IGI Global. doi:10.4018/978-1-5225-0700-0.ch001

Buse, C. G. (2017). Are Climate Change Adaptation Policies a Game Changer?: A Case Study of Perspectives from Public Health Officials in Ontario, Canada. In M. Bouzid (Ed.), *Examining the Role of Environmental Change on Emerging Infectious Diseases and Pandemics* (pp. 230–257). Hershey, PA: IGI Global. doi:10.4018/978-1-5225-0553-2.ch010

Calderon, F. A., Giolo, E. G., Frau, C. D., Rengel, M. G., Rodriguez, H., Tornello, M., ... Gallucci, R. (2018). Seismic Microzonation and Site Effects Detection Through Microtremors Measures: A Review. In N. Ceryan (Ed.), *Handbook of Research on Trends and Digital Advances in Engineering Geology* (pp. 326–349). Hershey, PA: IGI Global. doi:10.4018/978-1-5225-2709-1.ch009

Carfì, D., Donato, A., & Panuccio, D. (2018). A Game Theory Coopetitive Perspective for Sustainability of Global Feeding: Agreements Among Vegan and Non-Vegan Food Firms. In I. Management Association (Ed.), Game Theory: Breakthroughs in Research and Practice (pp. 71-104). Hershey, PA: IGI Global. https://doi.org/doi:10.4018/978-1-5225-2594-3.ch004

Castagnolo, V. (2018). Analyzing, Classifying, Safeguarding: Drawing for the Borgo Murattiano Neighbourhood of Bari. In G. Carlone, N. Martinelli, & F. Rotondo (Eds.), *Designing Grid Cities for Optimized Urban Development and Planning* (pp. 93–108). Hershey, PA: IGI Global. doi:10.4018/978-1-5225-3613-0.ch006

Chekima, B. (2018). The Dilemma of Purchase Intention: A Conceptual Framework for Understanding Actual Consumption of Organic Food. *International Journal of Sustainable Economies Management*, *7*(2), 1–13. doi:10.4018/IJSEM.2018040101

Chen, Y. (2017). Sustainable Supply Chains and International Soft Landings: A Case of Wetland Entrepreneurship. In B. Christiansen & F. Kasarcı (Eds.), *Corporate Espionage, Geopolitics, and Diplomacy Issues in International Business* (pp. 232–247). Hershey, PA: IGI Global. doi:10.4018/978-1-5225-1031-4.ch013

Çıtak, L., Akel, V., & Ersoy, E. (2018). Investors' Reactions to the Announcement of New Constituents of BIST Sustainability Index: An Analysis by Event Study and Mean-Median Tests. In M. Risso & S. Testarmata (Eds.), *Value Sharing for Sustainable and Inclusive Development* (pp. 270–289). Hershey, PA: IGI Global. doi:10.4018/978-1-5225-3147-0.ch012

D'Aleo, V., D'Aleo, F., & Bonanno, R. (2018). New Food Industries Toward a New Level of Sustainable Supply: Success Stories, Business Models, and Strategies. In V. Erokhin (Ed.), *Establishing Food Security and Alternatives to International Trade in Emerging Economies* (pp. 74–97). Hershey, PA: IGI Global. doi:10.4018/978-1-5225-2733-6.ch004

Dagevos, H., & Reinders, M. J. (2018). Flexitarianism and Social Marketing: Reflections on Eating Meat in Moderation. In D. Bogueva, D. Marinova, & T. Raphaely (Eds.), *Handbook of Research on Social Marketing and Its Influence on Animal Origin Food Product Consumption* (pp. 105–120). Hershey, PA: IGI Global. doi:10.4018/978-1-5225-4757-0.ch007

Danisman, G. O. (2022). What Drives Eco-Design Innovations in European SMEs? In U. Akkucuk (Ed.), *Disruptive Technologies and Eco-Innovation for Sustainable Development* (pp. 191–206). IGI Global. https://doi.org/10.4018/978-1-7998-8900-7.ch011

Deenapanray, P. N., & Ramma, I. (2017). Adaptations to Climate Change and Climate Variability in the Agriculture Sector in Mauritius: Lessons from a Technical Needs Assessment. In I. Management Association (Ed.), Natural Resources Management: Concepts, Methodologies, Tools, and Applications (pp. 655-680). Hershey, PA: IGI Global. https://doi.org/ doi:10.4018/978-1-5225-0803-8.ch030

Deenapanray, P. N., & Ramma, I. (2017). Adaptations to Climate Change and Climate Variability in the Agriculture Sector in Mauritius: Lessons from a Technical Needs Assessment. In I. Management Association (Ed.), Natural Resources Management: Concepts, Methodologies, Tools, and Applications (pp. 655-680). Hershey, PA: IGI Global. https://doi.org/ doi:10.4018/978-1-5225-0803-8.ch030

Deshpande, S., Basu, S. K., Li, X., & Chen, X. (2017). Smart, Innovative and Intelligent Technologies Used in Drug Designing. In I. Management Association (Ed.), Pharmaceutical Sciences: Breakthroughs in Research and Practice (pp. 1175-1191). Hershey, PA: IGI Global. https://doi.org/ doi:10.4018/978-1-5225-1762-7.ch045

Dlamini, P. N. (2017). Use of Information and Communication Technologies Tools to Capture, Store, and Disseminate Indigenous Knowledge: A Literature Review. In P. Ngulube (Ed.), *Handbook of Research on Theoretical Perspectives on Indigenous Knowledge Systems in Developing Countries* (pp. 225–247). Hershey, PA: IGI Global. doi:10.4018/978-1-5225-0833-5.ch010

Dolejsova, M., & Kera, D. (2017). The Fermentation GutHub Project and the Internet of Microbes. In S. Konomi & G. Roussos (Eds.), *Enriching Urban Spaces with Ambient Computing, the Internet of Things, and Smart City Design* (pp. 25–46). Hershey, PA: IGI Global. doi:10.4018/978-1-5225-0827-4.ch002

Dolunay, O. (2018). A Paradigm Shift: Empowering Farmers to Eliminate the Waste in the Form of Fresh Water and Energy Through the Implementation of 4R+T. In I. Management Association (Ed.), Sustainable Development: Concepts, Methodologies, Tools, and Applications (pp. 882-892). Hershey, PA: IGI Global. https://doi.org/doi:10.4018/978-1-5225-3817-2.ch039

Dube, P., Heijman, W. J., Ihle, R., & Ochieng, J. (2018). The Potential of Traditional Leafy Vegetables for Improving Food Security in Africa. In V. Erokhin (Ed.), *Establishing Food Security and Alternatives to International Trade in Emerging Economies* (pp. 220–243). Hershey, PA: IGI Global. doi:10.4018/978-1-5225-2733-6.ch011

Duruji, M. M., & Urenma, D. F. (2017). The Environmentalism and Politics of Climate Change: A Study of the Process of Global Convergence through UNFCCC Conferences. In I. Management Association (Ed.), *Natural Resources Management: Concepts, Methodologies, Tools, and Applications* (pp. 77-108). Hershey, PA: IGI Global. 10.4018/978-1-5225-0803-8.ch004

Dutta, U. (2017). Agro-Geoinformatics, Potato Cultivation, and Climate Change. In S. Londhe (Ed.), *Sustainable Potato Production and the Impact of Climate Change* (pp. 247–271). Hershey, PA: IGI Global. doi:10.4018/978-1-5225-1715-3.ch012

Edirisinghe, R., Stranieri, A., & Wickramasinghe, N. (2017). A Taxonomy for mHealth. In N. Wickramasinghe (Ed.), *Handbook of Research on Healthcare Administration and Management* (pp. 596–615). Hershey, PA: IGI Global. doi:10.4018/978-1-5225-0920-2.ch036

Ekpeni, N. M., & Ayeni, A. O. (2018). Global Natural Hazard and Disaster Vulnerability Management. In A. Eneanya (Ed.), *Handbook of Research on Environmental Policies for Emergency Management and Public Safety* (pp. 83–104). Hershey, PA: IGI Global. doi:10.4018/978-1-5225-3194-4.ch005

Ene, C., Voica, M. C., & Panait, M. (2017). Green Investments and Food Security: Opportunities and Future Directions in the Context of Sustainable Development. In M. Mieila (Ed.), *Measuring Sustainable Development and Green Investments in Contemporary Economies* (pp. 163–200). Hershey, PA: IGI Global. doi:10.4018/978-1-5225-2081-8.ch007

Escamilla, I., Ruíz, M. T., Ibarra, M. M., Soto, V. L., Quintero, R., & Guzmán, G. (2018). Geocoding Tweets Based on Semantic Web and Ontologies. In M. Lytras, N. Aljohani, E. Damiani, & K. Chui (Eds.), *Innovations, Developments, and Applications of Semantic Web and Information Systems* (pp. 372–392). Hershey, PA: IGI Global. doi:10.4018/978-1-5225-5042-6.ch014

Escribano, A. J. (2018). Marketing Strategies for Trendy Animal Products: Sustainability as a Core. In F. Quoquab, R. Thurasamy, & J. Mohammad (Eds.), *Driving Green Consumerism Through Strategic Sustainability Marketing* (pp. 169–203). Hershey, PA: IGI Global. doi:10.4018/978-1-5225-2912-5.ch010

Eudoxie, G., & Roopnarine, R. (2017). Climate Change Adaptation and Disaster Risk Management in the Caribbean. In W. Ganpat & W. Isaac (Eds.), *Environmental Sustainability and Climate Change Adaptation Strategies* (pp. 97–125). Hershey, PA: IGI Global. doi:10.4018/978-1-5225-1607-1.ch004

Farmer, L. S. (2017). Data Analytics for Strategic Management: Getting the Right Data. In V. Wang (Ed.), *Encyclopedia of Strategic Leadership and Management* (pp. 810–822). Hershey, PA: IGI Global. doi:10.4018/978-1-5225-1049-9.ch056

Fattal, L. R. (2017). Catastrophe: An Uncanny Catalyst for Creativity. In R. Shin (Ed.), *Convergence of Contemporary Art, Visual Culture, and Global Civic Engagement* (pp. 244–262). Hershey, PA: IGI Global. doi:10.4018/978-1-5225-1665-1.ch014

Forti, I. (2017). A Cross Reading of Landscape through Digital Landscape Models: The Case of Southern Garda. In A. Ippolito (Ed.), *Handbook of Research on Emerging Technologies for Architectural and Archaeological Heritage* (pp. 532–561). Hershey, PA: IGI Global. doi:10.4018/978-1-5225-0675-1.ch018

Gharbi, A., De Runz, C., & Akdag, H. (2017). Urban Development Modelling: A Survey. In S. Faiz & K. Mahmoudi (Eds.), *Handbook of Research on Geographic Information Systems Applications and Advancements* (pp. 96–124). Hershey, PA: IGI Global. doi:10.4018/978-1-5225-0937-0.ch004

Ghosh, I., & Ghoshal, I. (2018). Implications of Trade Liberalization for Food Security Under the ASEAN-India Strategic Partnership: A Gravity Model Approach. In V. Erokhin (Ed.), *Establishing Food Security and Alternatives to International Trade in Emerging Economies* (pp. 98–118). Hershey, PA: IGI Global. doi:10.4018/978-1-5225-2733-6.ch005

Glaves, H. M. (2017). Developing a Common Global Framework for Marine Data Management. In P. Diviacco, A. Leadbetter, & H. Glaves (Eds.), *Oceanographic and Marine Cross-Domain Data Management for Sustainable Development* (pp. 47–68). Hershey, PA: IGI Global. doi:10.4018/978-1-5225-0700-0.ch003

Godulla, A., & Wolf, C. (2018). Future of Food: Transmedia Strategies of National Geographic. In R. Gambarato & G. Alzamora (Eds.), *Exploring Transmedia Journalism in the Digital Age* (pp. 162–182). Hershey, PA: IGI Global. doi:10.4018/978-1-5225-3781-6.ch010

Gomes, P. P. (2018). Food and Environment: A Review on the Sustainability of Six Different Dietary Patterns. In A. Obayelu (Ed.), *Food Systems Sustainability and Environmental Policies in Modern Economies* (pp. 15–31). Hershey, PA: IGI Global. doi:10.4018/978-1-5225-3631-4.ch002

Gonzalez-Feliu, J. (2018). Sustainability Evaluation of Green Urban Logistics Systems: Literature Overview and Proposed Framework. In A. Paul, D. Bhattacharyya, & S. Anand (Eds.), *Green Initiatives for Business Sustainability and Value Creation* (pp. 103–134). Hershey, PA: IGI Global. doi:10.4018/978-1-5225-2662-9.ch005

Goodland, R. (2017). A Fresh Look at Livestock Greenhouse Gas Emissions and Mitigation. In I. Management Association (Ed.), Natural Resources Management: Concepts, Methodologies, Tools, and Applications (pp. 124-139). Hershey, PA: IGI Global. doi:10.4018/978-1-5225-0803-8.ch006

Goundar, S., & Appana, S. (2018). Mainstreaming Development Policies for Climate Change in Fiji: A Policy Gap Analysis and the Role of ICTs. In I. Management Association (Ed.), Sustainable Development: Concepts, Methodologies, Tools, and Applications (pp. 402-432). Hershey, PA: IGI Global. doi:10.4018/978-1-5225-3817-2.ch020

Granell-Canut, C., & Aguilar-Moreno, E. (2018). Geospatial Influence in Science Mapping. In M. Khosrow-Pour, D.B.A. (Ed.), Encyclopedia of Information Science and Technology, Fourth Edition (pp. 3473-3483). Hershey, PA: IGI Global. doi:10.4018/978-1-5225-2255-3.ch302

Grigelis, A., Blažauskas, N., Gelumbauskaitė, L. Ž., Gulbinskas, S., Suzdalev, S., & Ferrarin, C. (2017). Marine Environment Data Management Related to the Human Activity in the South-Eastern Baltic Sea (The Lithuanian Sector). In P. Diviacco, A. Leadbetter, & H. Glaves (Eds.), *Oceanographic and Marine Cross-Domain Data Management for Sustainable Development* (pp. 282–302). Hershey, PA: IGI Global. doi:10.4018/978-1-5225-0700-0.ch012

Guma, I. P., Rwashana, A. S., & Oyo, B. (2018). Food Security Policy Analysis Using System Dynamics: The Case of Uganda. *International Journal of Information Technologies and Systems Approach, 11*(1), 72–90. doi:10.4018/IJITSA.2018010104

Guma, I. P., Rwashana, A. S., & Oyo, B. (2018). Food Security Indicators for Subsistence Farmers Sustainability: A System Dynamics Approach. *International Journal of System Dynamics Applications, 7*(1), 45–64. doi:10.4018/IJSDA.2018010103

Gupta, P., & Goyal, S. (2017). Wildlife Habitat Evaluation. In A. Santra & S. Mitra (Eds.), *Remote Sensing Techniques and GIS Applications in Earth and Environmental Studies* (pp. 258–264). Hershey, PA: IGI Global. doi:10.4018/978-1-5225-1814-3.ch013

Hanson, T., & Hildebrand, E. (2018). GPS Travel Diaries in Rural Transportation Research: A Focus on Older Drivers. In I. Management Association (Ed.), Intelligent Transportation and Planning: Breakthroughs in Research and Practice (pp. 609-625). Hershey, PA: IGI Global. doi:10.4018/978-1-5225-5210-9.ch027

Hartman, M. B. (2017). Research-Based Climate Change Public Education Programs. In I. Management Association (Ed.), Natural Resources Management: Concepts, Methodologies, Tools, and Applications (pp. 992-1003). Hershey, PA: IGI Global. doi:10.4018/978-1-5225-0803-8.ch046

Hashim, N. (2017). Zanzibari Seaweed: Global Climate Change and the Promise of Adaptation. In I. Management Association (Ed.), Natural Resources Management: Concepts, Methodologies, Tools, and Applications (pp. 365-391). Hershey, PA: IGI Global. https://doi.org/ doi:10.4018/978-1-5225-0803-8.ch019

Herrera, J. E., Argüello, L. V., Gonzalez-Feliu, J., & Jaimes, W. A. (2017). Decision Support System Design Requirements, Information Management, and Urban Logistics Efficiency: Case Study of Bogotá, Colombia. In G. Jamil, A. Soares, & C. Pessoa (Eds.), *Handbook of Research on Information Management for Effective Logistics and Supply Chains* (pp. 223–238). Hershey, PA: IGI Global. doi:10.4018/978-1-5225-0973-8.ch012

Huizinga, T., Ayanso, A., Smoor, M., & Wronski, T. (2017). Exploring Insurance and Natural Disaster Tweets Using Text Analytics. *International Journal of Business Analytics, 4*(1), 1–17. doi:10.4018/IJBAN.2017010101

Hung, K., Kalantari, M., & Rajabifard, A. (2017). An Integrated Method for Assessing the Text Content Quality of Volunteered Geographic Information in Disaster Management. *International Journal of Information Systems for Crisis Response and Management, 9*(2), 1–17. doi:10.4018/IJISCRAM.2017040101

Husnain, A., & Avdic, A. (2018). Identifying the Contemporary Status of E-Service Sustainability Research. In I. Management Association (Ed.), Sustainable Development: Concepts, Methodologies, Tools, and Applications (pp. 467-485). Hershey, PA: IGI Global. https://doi.org/ doi:10.4018/978-1-5225-3817-2.ch022

Iarossi, M. P., & Ferro, L. (2017). "The Past is Never Dead. It's Not Even Past": Virtual Archaeological Promenade. In A. Ippolito & M. Cigola (Eds.), *Handbook of Research on Emerging Technologies for Digital Preservation and Information Modeling* (pp. 228–255). Hershey, PA: IGI Global. doi:10.4018/978-1-5225-0680-5.ch010

Ignjatijević, S., & Cvijanović, D. (2018). Analysis of Serbian Honey Production and Exports. In *Exploring the Global Competitiveness of Agri-Food Sectors and Serbia's Dominant Presence: Emerging Research and Opportunities* (pp. 109–139). Hershey, PA: IGI Global. doi:10.4018/978-1-5225-2762-6.ch005

Jana, S. K., & Karmakar, A. K. (2017). Globalization, Governance, and Food Security: The Case of BRICS. In I. Management Association (Ed.), Natural Resources Management: Concepts, Methodologies, Tools, and Applications (pp. 692-712). Hershey, PA: IGI Global. https://doi.org/ doi:10.4018/978-1-5225-0803-8.ch032

Jana, S. K., & Karmakar, A. K. (2017). Food Security in Asia: Is There Convergence? In I. Management Association (Ed.), Natural Resources Management: Concepts, Methodologies, Tools, and Applications (pp. 109-123). Hershey, PA: IGI Global. https://doi.org/ doi:10.4018/978-1-5225-0803-8.ch005

John, J., & Kumar, S. (2018). A Locational Decision Making Framework for Shipbreaking Under Multiple Criteria. In I. Management Association (Ed.), Operations and Service Management: Concepts, Methodologies, Tools, and Applications (pp. 504-527). Hershey, PA: IGI Global. doi:10.4018/978-1-5225-3909-4.ch024

John, J., & Srivastava, R. K. (2018). Decision Insights for Shipbreaking using Environmental Impact Assessment: Review and Perspectives. *International Journal of Strategic Decision Sciences, 9*(1), 45–62. doi:10.4018/IJSDS.2018010104

Joshi, Y., & Rahman, Z. (2018). Determinants of Sustainable Consumption Behaviour: Review and Conceptual Framework. In A. Paul, D. Bhattacharyya, & S. Anand (Eds.), *Green Initiatives for Business Sustainability and Value Creation* (pp. 239–262). Hershey, PA: IGI Global. doi:10.4018/978-1-5225-2662-9.ch011

Juma, D. W., Reuben, M., Wang, H., & Li, F. (2018). Adaptive Coevolution: Realigning the Water Governance Regime to the Changing Climate. In I. Management Association (Ed.), Hydrology and Water Resource Management: Breakthroughs in Research and Practice (pp. 346-357). Hershey, PA: IGI Global. https://doi.org/ doi:10.4018/978-1-5225-3427-3.ch014

K., S., & Tripathy, B. K. (2018). Neighborhood Rough-Sets-Based Spatial Data Analytics. In M. Khosrow-Pour, D.B.A. (Ed.), *Encyclopedia of Information Science and Technology, Fourth Edition* (pp. 1835-1844). Hershey, PA: IGI Global. https:// doi.org/ doi:10.4018/978-1-5225-2255-3.ch160

Kabir, F. (2018). Towards a More Gender-Inclusive Climate Change Policy. In N. Mahtab, T. Haque, I. Khan, M. Islam, & I. Wahid (Eds.), *Handbook of Research on Women's Issues and Rights in the Developing World* (pp. 354–369). Hershey, PA: IGI Global. doi:10.4018/978-1-5225-3018-3.ch022

Kabir, F. (2018). Towards a More Gender-Inclusive Climate Change Policy. In I. Management Association (Ed.), Climate Change and Environmental Concerns: Breakthroughs in Research and Practice (pp. 525-540). Hershey, PA: IGI Global. https://doi.org/ doi:10.4018/978-1-5225-5487-5.ch027

Kanyamuka, J. S., Jumbe, C. B., & Ricker-Gilbert, J. (2018). Making Agricultural Input Subsidies More Effective and Profitable in Africa: The Role of Complementary Interventions. In A. Obayelu (Ed.), *Food Systems Sustainability and Environmental Policies in Modern Economies* (pp. 172–187). Hershey, PA: IGI Global. doi:10.4018/978-1-5225-3631-4.ch008

Karimi, H., & Gholamrezafahimi, F. (2017). Study of Integrated Coastal Zone Management and Its Environmental Effects: A Case of Iran. In R. Singh, A. Singh, & V. Srivastava (Eds.), *Environmental Issues Surrounding Human Overpopulation* (pp. 64–88). Hershey, PA: IGI Global. doi:10.4018/978-1-5225-1683-5.ch004

Kaya, I. R., Hutabarat, J., & Bambang, A. N. (2018). "Sasi": A New Path to Sustain Seaweed Farming From Up-Stream to Down-Stream in Kotania Bay, Molucass. *International Journal of Social Ecology and Sustainable Development, 9*(2), 28–36. doi:10.4018/IJSESD.2018040103

Khader, V. (2018). Technologies for Food, Health, Livelihood, and Nutrition Security. In I. Management Association (Ed.), Food Science and Nutrition: Breakthroughs in Research and Practice (pp. 94-112). Hershey, PA: IGI Global. https://doi.org/doi:10.4018/978-1-5225-5207-9.ch005

Kocadağlı, A. Y. (2017). The Temporal and Spatial Development of Organic Agriculture in Turkey. In W. Ganpat, R. Dyer, & W. Isaac (Eds.), *Agricultural Development and Food Security in Developing Nations* (pp. 130–156). Hershey, PA: IGI Global. doi:10.4018/978-1-5225-0942-4.ch006

Koundouri, P., Giannouli, A., & Souliotis, I. (2017). An Integrated Approach for Sustainable Environmental and Socio-Economic Development Using Offshore Infrastructure. In I. Management Association (Ed.), Renewable and Alternative Energy: Concepts, Methodologies, Tools, and Applications (pp. 1581-1601). Hershey, PA: IGI Global. doi:10.4018/978-1-5225-1671-2.ch056

Kumar, A., & Dash, M. K. (2017). Sustainability and Future Generation Infrastructure on Digital Platform: A Study of Generation Y. In N. Ray (Ed.), *Business Infrastructure for Sustainability in Developing Economies* (pp. 124–142). Hershey, PA: IGI Global. doi:10.4018/978-1-5225-2041-2.ch007

Kumar, A., Mukherjee, A. B., & Krishna, A. P. (2017). Application of Conventional Data Mining Techniques and Web Mining to Aid Disaster Management. In A. Kumar (Ed.), *Web Usage Mining Techniques and Applications Across Industries* (pp. 138–167). Hershey, PA: IGI Global. doi:10.4018/978-1-5225-0613-3.ch006

Kumar, C. P. (2017). Impact of Climate Change on Groundwater Resources. In I. Management Association (Ed.), Natural Resources Management: Concepts, Methodologies, Tools, and Applications (pp. 1094-1120). Hershey, PA: IGI Global. doi:10.4018/978-1-5225-0803-8.ch052

Kumari, S., & Patil, Y. (2017). Achieving Climate Smart Agriculture with a Sustainable Use of Water: A Conceptual Framework for Sustaining the Use of Water for Agriculture in the Era of Climate Change. In P. Rao & Y. Patil (Eds.), *Reconsidering the Impact of Climate Change on Global Water Supply, Use, and Management* (pp. 122–143). Hershey, PA: IGI Global. doi:10.4018/978-1-5225-1046-8.ch008

Kumari, S., & Patil, Y. (2018). Achieving Climate Smart Agriculture With a Sustainable Use of Water: A Conceptual Framework for Sustaining the Use of Water for Agriculture in the Era of Climate Change. In I. Management Association (Ed.), Climate Change and Environmental Concerns: Breakthroughs in Research and Practice (pp. 111-133). Hershey, PA: IGI Global. doi:10.4018/978-1-5225-5487-5.ch006

Kursah, M. B. (2017). Least-Cost Pipeline using Geographic Information System: The Limit to Technicalities. *International Journal of Applied Geospatial Research, 8*(3), 1–15. doi:10.4018/ijagr.2017070101

Lahiri, S., Ghosh, D., & Bhakta, J. N. (2017). Role of Microbes in Eco-Remediation of Perturbed Aquatic Ecosystem. In J. Bhakta (Ed.), *Handbook of Research on Inventive Bioremediation Techniques* (pp. 70–107). Hershey, PA: IGI Global. doi:10.4018/978-1-5225-2325-3.ch004

Lallo, C. H., Smalling, S., Facey, A., & Hughes, M. (2018). The Impact of Climate Change on Small Ruminant Performance in Caribbean Communities. In I. Management Association (Ed.), Climate Change and Environmental Concerns: Breakthroughs in Research and Practice (pp. 193-218). Hershey, PA: IGI Global. doi:10.4018/978-1-5225-5487-5.ch010

Laurini, R. (2017). Nature of Geographic Knowledge Bases. In S. Faiz & K. Mahmoudi (Eds.), *Handbook of Research on Geographic Information Systems Applications and Advancements* (pp. 29–60). Hershey, PA: IGI Global. doi:10.4018/978-1-5225-0937-0.ch002

Lawrence, J., Simpson, L., & Piggott, A. (2017). Protected Agriculture: A Climate Change Adaptation for Food and Nutrition Security. In I. Management Association (Ed.), Natural Resources Management: Concepts, Methodologies, Tools, and Applications (pp. 140-158). Hershey, PA: IGI Global. doi:10.4018/978-1-5225-0803-8.ch007

Leadbetter, A., Cheatham, M., Shepherd, A., & Thomas, R. (2017). Linked Ocean Data 2.0. In P. Diviacco, A. Leadbetter, & H. Glaves (Eds.), *Oceanographic and Marine Cross-Domain Data Management for Sustainable Development* (pp. 69–99). Hershey, PA: IGI Global. doi:10.4018/978-1-5225-0700-0.ch004

Lucas, M. R., Rego, C., Vieira, C., & Vieira, I. (2017). Proximity and Cooperation for Innovative Regional Development: The Case of the Science and Technology Park of Alentejo. In L. Carvalho (Ed.), *Handbook of Research on Entrepreneurial Development and Innovation Within Smart Cities* (pp. 199–228). Hershey, PA: IGI Global. doi:10.4018/978-1-5225-1978-2.ch010

Ma, X., Beaulieu, S. E., Fu, L., Fox, P., Di Stefano, M., & West, P. (2017). Documenting Provenance for Reproducible Marine Ecosystem Assessment in Open Science. In P. Diviacco, A. Leadbetter, & H. Glaves (Eds.), *Oceanographic and Marine Cross-Domain Data Management for Sustainable Development* (pp. 100–126). Hershey, PA: IGI Global. doi:10.4018/978-1-5225-0700-0.ch005

Mabe, L. K., & Oladele, O. I. (2017). Application of Information Communication Technologies for Agricultural Development through Extension Services: A Review. In T. Tossy (Ed.), *Information Technology Integration for Socio-Economic Development* (pp. 52–101). Hershey, PA: IGI Global. doi:10.4018/978-1-5225-0539-6.ch003

Malomo, B. I. (2018). A Review of Psychological Resilience as a Response to Natural Hazards in Nigeria. In A. Eneanya (Ed.), *Handbook of Research on Environmental Policies for Emergency Management and Public Safety* (pp. 147–165). Hershey, PA: IGI Global. doi:10.4018/978-1-5225-3194-4.ch008

Manchiraju, S. (2018). Predicting Behavioral Intentions Toward Sustainable Fashion Consumption: A Comparison of Attitude-Behavior and Value-Behavior Consistency Models. In I. Management Association (Ed.), Fashion and Textiles: Breakthroughs in Research and Practice (pp. 1-21). Hershey, PA: IGI Global. doi:10.4018/978-1-5225-3432-7.ch001

Manzella, G. M., Bartolini, R., Bustaffa, F., D'Angelo, P., De Mattei, M., Frontini, F., ... Spada, A. (2017). Semantic Search Engine for Data Management and Sustainable Development: Marine Planning Service Platform. In P. Diviacco, A. Leadbetter, & H. Glaves (Eds.), *Oceanographic and Marine Cross-Domain Data Management for Sustainable Development* (pp. 127–154). Hershey, PA: IGI Global. doi:10.4018/978-1-5225-0700-0.ch006

Mbonigaba, J. (2018). Comparing the Effects of Unsustainable Production and Consumption of Food on Health and Policy Across Developed and Less Developed Countries. In A. Obayelu (Ed.), *Food Systems Sustainability and Environmental Policies in Modern Economies* (pp. 124–158). Hershey, PA: IGI Global. doi:10.4018/978-1-5225-3631-4.ch006

McKeown, A. E. (2017). Nurses, Healthcare, and Environmental Pollution and Solutions: Breaking the Cycle of Harm. In I. Management Association (Ed.), Natural Resources Management: Concepts, Methodologies, Tools, and Applications (pp. 392-415). Hershey, PA: IGI Global. https://doi.org/ doi:10.4018/978-1-5225-0803-8.ch020

Mili, B., Barua, A., & Katyaini, S. (2017). Climate Change and Adaptation through the Lens of Capability Approach: A Case Study from Darjeeling, Eastern Himalaya. In I. Management Association (Ed.), Natural Resources Management: Concepts, Methodologies, Tools, and Applications (pp. 1351-1365). Hershey, PA: IGI Global. https://doi.org/ doi:10.4018/978-1-5225-0803-8.ch064

Mir, S. A., Shah, M. A., Mir, M. M., & Iqbal, U. (2017). New Horizons of Nanotechnology in Agriculture and Food Processing Industry. In B. Nayak, A. Nanda, & M. Bhat (Eds.), *Integrating Biologically-Inspired Nanotechnology into Medical Practice* (pp. 230–258). Hershey, PA: IGI Global. doi:10.4018/978-1-5225-0610-2.ch009

Moallem, M., Sterrett, W. L., Gordon, C. R., Sukhera, S. M., Mahmood, A., & Bashir, A. (2021). An Investigation of the Effects of Integrating Computing and Project- or Problem-Based Learning in the Context of Environmental Sciences: A Case of Pakistani STEM Teachers. In S. Schroth & J. Daniels (Eds.), *Building STEM Skills Through Environmental Education* (pp. 49–89). IGI Global. https://doi.org/10.4018/978-1-7998-2711-5.ch003

Mujere, N., & Moyce, W. (2017). Climate Change Impacts on Surface Water Quality. In W. Ganpat & W. Isaac (Eds.), *Environmental Sustainability and Climate Change Adaptation Strategies* (pp. 322–340). Hershey, PA: IGI Global. doi:10.4018/978-1-5225-1607-1.ch012

Mukherjee, A. B., Krishna, A. P., & Patel, N. (2018). Geospatial Technology for Urban Sciences. In *Geospatial Technologies in Urban System Development: Emerging Research and Opportunities* (pp. 99–120). Hershey, PA: IGI Global. doi:10.4018/978-1-5225-3683-3.ch005

Nagarajan, S. K., & Sangaiah, A. K. (2017). Vegetation Index: Ideas, Methods, Influences, and Trends. In N. Kumar, A. Sangaiah, M. Arun, & S. Anand (Eds.), *Advanced Image Processing Techniques and Applications* (pp. 347–386). Hershey, PA: IGI Global. doi:10.4018/978-1-5225-2053-5.ch016

Naraine, L., & Meehan, K. (2017). Strengthening Food Security with Sustainable Practices by Smallholder Farmers in Lesser Developed Economies. In W. Ganpat, R. Dyer, & W. Isaac (Eds.), *Agricultural Development and Food Security in Developing Nations* (pp. 57–81). Hershey, PA: IGI Global. doi:10.4018/978-1-5225-0942-4.ch003

Naraine, L., & Meehan, K. (2017). Strengthening Food Security with Sustainable Practices by Smallholder Farmers in Lesser Developed Economies. In W. Ganpat, R. Dyer, & W. Isaac (Eds.), *Agricultural Development and Food Security in Developing Nations* (pp. 57–81). Hershey, PA: IGI Global. doi:10.4018/978-1-5225-0942-4.ch003

Naraine, L., & Meehan, K. (2017). Strengthening Food Security with Sustainable Practices by Smallholder Farmers in Lesser Developed Economies. In W. Ganpat, R. Dyer, & W. Isaac (Eds.), *Agricultural Development and Food Security in Developing Nations* (pp. 57–81). Hershey, PA: IGI Global. doi:10.4018/978-1-5225-0942-4.ch003

Nikolaou, K., Tsakiridou, E., Anastasiadis, F., & Mattas, K. (2017). Exploring Alternative Distribution Channels of Agricultural Products. *International Journal of Food and Beverage Manufacturing and Business Models*, *2*(2), 36–66. doi:10.4018/IJFBMBM.2017070103

Nishat, K. J., & Rahman, M. S. (2018). Disaster, Vulnerability, and Violence Against Women: Global Findings and a Research Agenda for Bangladesh. In N. Mahtab, T. Haque, I. Khan, M. Islam, & I. Wahid (Eds.), *Handbook of Research on Women's Issues and Rights in the Developing World* (pp. 235–250). Hershey, PA: IGI Global. doi:10.4018/978-1-5225-3018-3.ch014

O'Hara, S., Jones, D., & Trobman, H. B. (2018). Building an Urban Food System Through UDC Food Hubs. In A. Burtin, J. Fleming, & P. Hampton-Garland (Eds.), *Changing Urban Landscapes Through Public Higher Education* (pp. 116–143). Hershey, PA: IGI Global. doi:10.4018/978-1-5225-3454-9.ch006

Obayelu, A. E. (2018). Integrating Environment, Food Systems, and Sustainability in Feeding the Growing Population in Developing Countries. In A. Obayelu (Ed.), *Food Systems Sustainability and Environmental Policies in Modern Economies* (pp. 1–14). Hershey, PA: IGI Global. doi:10.4018/978-1-5225-3631-4.ch001

Othman, R., Nath, N., & Laswad, F. (2018). Environmental Reporting and Accounting: Sustainability Hybridisation. In G. Azevedo, J. da Silva Oliveira, R. Marques, & A. Ferreira (Eds.), *Handbook of Research on Modernization and Accountability in Public Sector Management* (pp. 130–158). Hershey, PA: IGI Global. doi:10.4018/978-1-5225-3731-1.ch007

Ouadi, A., & Zitouni, A. (2021). Phasor Measurement Improvement Using Digital Filter in a Smart Grid. In A. Recioui & H. Bentarzi (Eds.), *Optimizing and Measuring Smart Grid Operation and Control* (pp. 100–117). IGI Global. https://doi.org/10.4018/978-1-7998-4027-5.ch005

Padigala, B. S. (2018). Traditional Water Management System for Climate Change Adaptation in Mountain Ecosystems. In I. Management Association (Ed.), Climate Change and Environmental Concerns: Breakthroughs in Research and Practice (pp. 630-655). Hershey, PA: IGI Global. doi:10.4018/978-1-5225-5487-5.ch033

Panda, C. K. (2018). Mobile Phone Usage in Agricultural Extension in India: The Current and Future Perspective. In F. Mtenzi, G. Oreku, D. Lupiana, & J. Yonazi (Eds.), *Mobile Technologies and Socio-Economic Development in Emerging Nations* (pp. 1–21). Hershey, PA: IGI Global. doi:10.4018/978-1-5225-4029-8.ch001

Pandian, S. L., Yarrakula, K., & Chaudhury, P. (2018). GIS-Based Decision Support System for Village Level: A Case Study in Andhra Pradesh. In S. Chaudhuri & N. Ray (Eds.), *GIS Applications in the Tourism and Hospitality Industry* (pp. 275–295). Hershey, PA: IGI Global. doi:10.4018/978-1-5225-5088-4.ch012

Quaranta, G., & Salvia, R. (2017). Social-Based Product Innovation and Governance in The Milk Sector: The Case of Carciocacio and Innonatura. In T. Tarnanidis, M. Vlachopoulou, & J. Papathanasiou (Eds.), *Driving Agribusiness With Technology Innovations* (pp. 293–310). Hershey, PA: IGI Global. doi:10.4018/978-1-5225-2107-5.ch015

Rahman, M. K., Schmidlin, T. W., Munro-Stasiuk, M. J., & Curtis, A. (2017). Geospatial Analysis of Land Loss, Land Cover Change, and Landuse Patterns of Kutubdia Island, Bangladesh. *International Journal of Applied Geospatial Research*, *8*(2), 45–60. doi:10.4018/IJAGR.2017040104

Rajack-Talley, T. A. (2017). Agriculture, Trade Liberalization and Poverty in the ACP Countries. In W. Ganpat, R. Dyer, & W. Isaac (Eds.), *Agricultural Development and Food Security in Developing Nations* (pp. 1–20). Hershey, PA: IGI Global. doi:10.4018/978-1-5225-0942-4.ch001

Rajamanickam, S. (2018). Exploring Landscapes in Regional Convergence: Environment and Sustainable Development in South Asia. In I. Management Association (Ed.), Sustainable Development: Concepts, Methodologies, Tools, and Applications (pp. 1051-1087). Hershey, PA: IGI Global. https://doi.org/ doi:10.4018/978-1-5225-3817-2.ch047

Rizvi, S. M., & Dearden, A. (2018). KHETI: ICT Solution for Agriculture Extension and Its Replication in Open and Distance Learning. In U. Pandey & V. Indrakanti (Eds.), *Open and Distance Learning Initiatives for Sustainable Development* (pp. 163–174). Hershey, PA: IGI Global. doi:10.4018/978-1-5225-2621-6.ch008

Roşu, L., & Macarov, L. I. (2017). Management of Drought and Floods in the Dobrogea Region. In I. Management Association (Ed.), Agri-Food Supply Chain Management: Breakthroughs in Research and Practice (pp. 372-403). Hershey, PA: IGI Global. https://doi.org/ doi:10.4018/978-1-5225-1629-3.ch016

Rouzbehani, K., & Rouzbehani, S. (2018). Mapping Women's World: GIS and the Case of Breast Cancer in the US. *International Journal of Public Health Management and Ethics*, *3*(1), 14–25. doi:10.4018/IJPHME.2018010102

Roy, D. (2018). Success Factors of Adoption of Mobile Applications in Rural India: Effect of Service Characteristics on Conceptual Model. In M. Khosrow-Pour, D.B.A. (Ed.), Green Computing Strategies for Competitive Advantage and Business Sustainability (pp. 211-238). Hershey, PA: IGI Global. https://doi.org/ doi:10.4018/978-1-5225-5017-4.ch010

Sajeva, M., Lemon, M., & Sahota, P. S. (2017). Governance for Food Security: A Framework for Social Learning and Scenario Building. *International Journal of Food and Beverage Manufacturing and Business Models*, *2*(2), 67–84. doi:10.4018/IJFBMBM.2017070104

Sambhanthan, A., & Potdar, V. (2017). A Study of the Parameters Impacting Sustainability in Information Technology Organizations. *International Journal of Knowledge-Based Organizations*, *7*(3), 27–39. doi:10.4018/IJKBO.2017070103

Sanga, C., Kalungwizi, V. J., & Msuya, C. P. (2017). Bridging Gender Gaps in Provision of Agricultural Extension Service Using ICT: Experiences from Sokoine University of Agriculture (SUA) Farmer Voice Radio (FVR) Project in Tanzania. In I. Management Association (Ed.), Discrimination and Diversity: Concepts, Methodologies, Tools, and Applications (pp. 682-697). Hershey, PA: IGI Global. https://doi.org/ doi:10.4018/978-1-5225-1933-1.ch031

Santra, A., & Mitra, D. (2017). Role of Remote Sensing in Potential Fishing Zone Forecast. In A. Santra & S. Mitra (Eds.), *Remote Sensing Techniques and GIS Applications in Earth and Environmental Studies* (pp. 243–257). Hershey, PA: IGI Global. doi:10.4018/978-1-5225-1814-3.ch012

Schaap, D. (2017). SeaDataNet: Towards a Pan-European Infrastructure for Marine and Ocean Data Management. In P. Diviacco, A. Leadbetter, & H. Glaves (Eds.), *Oceanographic and Marine Cross-Domain Data Management for Sustainable Development* (pp. 155–177). Hershey, PA: IGI Global. doi:10.4018/978-1-5225-0700-0.ch007

Seckin-Celik, T. (2017). Sustainability Reporting and Sustainability in the Turkish Business Context. In U. Akkucuk (Ed.), *Ethics and Sustainability in Global Supply Chain Management* (pp. 115–132). Hershey, PA: IGI Global. doi:10.4018/978-1-5225-2036-8.ch006

Segbefia, A. Y., Barnes, V. R., Akpalu, L. A., & Mensah, M. (2018). Environmental Location Assessment for Seaweed Cultivation in Ghana: A Spatial Multi-Criteria Approach. *International Journal of Applied Geospatial Research*, *9*(1), 51–64. doi:10.4018/IJAGR.2018010104

Sen, Y. (2018). How to Manage Sustainability: A Framework for Corporate Sustainability Tools. In I. Management Association (Ed.), Sustainable Development: Concepts, Methodologies, Tools, and Applications (pp. 568-589). Hershey, PA: IGI Global. doi:10.4018/978-1-5225-3817-2.ch026

Shamshiry, E., Abdulai, A. M., Mokhtar, M. B., & Komoo, I. (2015). Regional Landfill Site Selection with GIS and Analytical Hierarchy Process Techniques: A Case Study of Langkawi Island, Malaysia. In P. Thomas, M. Srihari, & S. Kaur (Eds.), *Handbook of Research on Cultural and Economic Impacts of the Information Society* (pp. 248–282). Hershey, PA: IGI Global. doi:10.4018/978-1-4666-8598-7.ch011

Sharma, Y. K., Mangla, S. K., Patil, P. P., & Uniyal, S. (2018). Analyzing Sustainable Food Supply Chain Management Challenges in India. In M. Ram & J. Davim (Eds.), *Soft Computing Techniques and Applications in Mechanical Engineering* (pp. 162–180). Hershey, PA: IGI Global. doi:10.4018/978-1-5225-3035-0.ch008

Silvestrelli, P. (2018). The Impact of Events: To Which Extent Are They Sustainable for Tourist Destinations? Some Evidences From Expo Milano 2015. In M. Risso & S. Testarmata (Eds.), *Value Sharing for Sustainable and Inclusive Development* (pp. 185–204). Hershey, PA: IGI Global. doi:10.4018/978-1-5225-3147-0.ch008

Silvius, G. (2018). Sustainability Evaluation of IT/IS Projects. In I. Management Association (Ed.), Sustainable Development: Concepts, Methodologies, Tools, and Applications (pp. 26-40). Hershey, PA: IGI Global. https://doi.org/ doi:10.4018/978-1-5225-3817-2.ch002

Singh, R., Srivastava, P., Singh, P., Upadhyay, S., & Raghubanshi, A. S. (2017). Human Overpopulation and Food Security: Challenges for the Agriculture Sustainability. In R. Singh, A. Singh, & V. Srivastava (Eds.), *Environmental Issues Surrounding Human Overpopulation* (pp. 12–39). Hershey, PA: IGI Global. doi:10.4018/978-1-5225-1683-5.ch002

Singh, R., Srivastava, P., Singh, P., Upadhyay, S., & Raghubanshi, A. S. (2017). Human Overpopulation and Food Security: Challenges for the Agriculture Sustainability. In R. Singh, A. Singh, & V. Srivastava (Eds.), *Environmental Issues Surrounding Human Overpopulation* (pp. 12–39). Hershey, PA: IGI Global. doi:10.4018/978-1-5225-1683-5.ch002

Srivastava, N. (2017). Climate Change Mitigation: Collective Efforts and Responsibly. In I. Management Association (Ed.), Natural Resources Management: Concepts, Methodologies, Tools, and Applications (pp. 64-76). Hershey, PA: IGI Global. https://doi.org/ doi:10.4018/978-1-5225-0803-8.ch003

Stanganelli, M., & Gerundo, C. (2017). Understanding the Role of Urban Morphology and Green Areas Configuration During Heat Waves. *International Journal of Agricultural and Environmental Information Systems*, *8*(2), 50–64. doi:10.4018/IJAEIS.2017040104

Stewart, M. K., Hagood, D., & Ching, C. C. (2017). Virtual Games and Real-World Communities: Environments that Constrain and Enable Physical Activity in Games for Health. *International Journal of Game-Based Learning*, *7*(1), 1–19. doi:10.4018/IJGBL.2017010101

Stone, R. J. (2017). Modelling the Frequency of Tropical Cyclones in the Lower Caribbean Region. In W. Ganpat & W. Isaac (Eds.), *Environmental Sustainability and Climate Change Adaptation Strategies* (pp. 341–349). Hershey, PA: IGI Global. doi:10.4018/978-1-5225-1607-1.ch013

Syed, A., & Jabeen, U. A. (2018). Climate Change Impact on Agriculture and Food Security. In A. Eneanya (Ed.), *Handbook of Research on Environmental Policies for Emergency Management and Public Safety* (pp. 223–237). Hershey, PA: IGI Global. doi:10.4018/978-1-5225-3194-4.ch012

Tam, G. C. (2017). The Global View of Sustainability. In *Managerial Strategies and Green Solutions for Project Sustainability* (pp. 1–24). Hershey, PA: IGI Global. doi:10.4018/978-1-5225-2371-0.ch001

Tam, G. C. (2017). Understanding Project Sustainability. In *Managerial Strategies and Green Solutions for Project Sustainability* (pp. 110–139). Hershey, PA: IGI Global. doi:10.4018/978-1-5225-2371-0.ch005

Tam, G. C. (2017). Perspectives on Sustainability. In *Managerial Strategies and Green Solutions for Project Sustainability* (pp. 53–76). Hershey, PA: IGI Global. doi:10.4018/978-1-5225-2371-0.ch003

Tang, M., & Karunanithi, A. T. (2018). Visual Logic Maps (vLms). In *Advanced Concept Maps in STEM Education: Emerging Research and Opportunities* (pp. 108–149). Hershey, PA: IGI Global. doi:10.4018/978-1-5225-2184-6.ch005

Taşçıoğlu, M., & Yener, D. (2018). The Value and Scope of GIS in Marketing and Tourism Management. In S. Chaudhuri & N. Ray (Eds.), *GIS Applications in the Tourism and Hospitality Industry* (pp. 189–211). Hershey, PA: IGI Global. doi:10.4018/978-1-5225-5088-4.ch009

Thanh Tung, B. (2021). Pharmacology and Therapeutic Applications of Resveratrol. In A. Hussain & S. Behl (Eds.), *Treating Endocrine and Metabolic Disorders With Herbal Medicines* (pp. 321–333). IGI Global. https://doi.org/10.4018/978-1-7998-4808-0.ch014

Tianming, G. (2018). Food Security and Rural Development on Emerging Markets of Northeast Asia: Cases of Chinese North and Russian Far East. In V. Erokhin (Ed.), *Establishing Food Security and Alternatives to International Trade in Emerging Economies* (pp. 155–176). Hershey, PA: IGI Global. doi:10.4018/978-1-5225-2733-6.ch008

Tiftikçigil, B. Y., Yaşgül, Y. S., & Güriş, B. (2017). Sustainability of Foreign Trade Deficit in Energy: The Case of Turkey. In N. Ray (Ed.), *Business Infrastructure for Sustainability in Developing Economies* (pp. 94–109). Hershey, PA: IGI Global. doi:10.4018/978-1-5225-2041-2.ch005

Tiwari, S., Vaish, B., & Singh, P. (2018). Population and Global Food Security: Issues Related to Climate Change. In I. Management Association (Ed.), Climate Change and Environmental Concerns: Breakthroughs in Research and Practice (pp. 41-64). Hershey, PA: IGI Global. https://doi.org/ doi:10.4018/978-1-5225-5487-5.ch003

Toujani, A., & Achour, H. (2018). A Data Mining Framework for Forest Fire Mapping. In I. Management Association (Ed.), Information Retrieval and Management: Concepts, Methodologies, Tools, and Applications (pp. 771-794). Hershey, PA: IGI Global. doi:10.4018/978-1-5225-5191-1.ch033

Tripathy, B., & K., S. B. (2017). Rough Fuzzy Set Theory and Neighbourhood Approximation Based Modelling for Spatial Epidemiology. In I. Management Association (Ed.), *Public Health and Welfare: Concepts, Methodologies, Tools, and Applications* (pp. 1257-1268). Hershey, PA: IGI Global. https://doi.org/ doi:10.4018/978-1-5225-1674-3.ch058

Trukhachev, A. (2017). New Approaches to Regional Branding through Green Production and Utilization of Existing Natural Advantages. In I. Management Association (Ed.), Advertising and Branding: Concepts, Methodologies, Tools, and Applications (pp. 1758-1778). Hershey, PA: IGI Global. doi:10.4018/978-1-5225-1793-1.ch081

Tsobanoglou, G. O., & Vlachopoulou, E. I. (2017). Social-Ecological Systems in Local Fisheries Communities. In G. Korres, E. Kourliouros, & M. Michailidis (Eds.), *Handbook of Research on Policies and Practices for Sustainable Economic Growth and Regional Development* (pp. 306–316). Hershey, PA: IGI Global. doi:10.4018/978-1-5225-2458-8.ch026

Tuydes-Yaman, H., & Karatas, P. (2018). Evaluation of Walkability and Pedestrian Level of Service. In I. Management Association (Ed.), Intelligent Transportation and Planning: Breakthroughs in Research and Practice (pp. 264-291). Hershey, PA: IGI Global. doi:10.4018/978-1-5225-5210-9.ch012

Uddin, S., Chakravorty, S., Ray, A., & Sherpa, K. S. (2018). Optimal Location of Sub-Station Using Q-GIS and Multi-Criteria Decision Making Approach. *International Journal of Decision Support System Technology*, *10*(2), 65–79. doi:10.4018/IJDSST.2018040104

Uddin, S., Chakravorty, S., Sherpa, K. S., & Ray, A. (2018). Power Distribution System Planning Using Q-GIS. *International Journal of Energy Optimization and Engineering*, *7*(2), 61–75. doi:10.4018/IJEOE.2018040103

Uzun, F. V. (2018). Natural Resources Management. In A. Eneanya (Ed.), *Handbook of Research on Environmental Policies for Emergency Management and Public Safety* (pp. 1–21). Hershey, PA: IGI Global. doi:10.4018/978-1-5225-3194-4.ch001

V, M., Agrawal, R., Sharma, V., & T.N., K. (2018). Supply Chain Social Sustainability and Manufacturing. In I. Management Association (Ed.), *Technology Adoption and Social Issues: Concepts, Methodologies, Tools, and Applications* (pp. 226-252). Hershey, PA: IGI Global. https://doi.org/ doi:10.4018/978-1-5225-5201-7.ch011

van der Vliet-Bakker, J. M. (2017). Environmentally Forced Migration and Human Rights. In C. Akrivopoulou (Ed.), *Defending Human Rights and Democracy in the Era of Globalization* (pp. 146–180). Hershey, PA: IGI Global. doi:10.4018/978-1-5225-0723-9.ch007

Vargas-Hernández, J. G. (2022). A Comprehensive Entrepreneurship Model for the Internationalization of Green Innovation Businesses. In U. Akkucuk (Ed.), *Disruptive Technologies and Eco-Innovation for Sustainable Development* (pp. 131–149). IGI Global. https://doi.org/10.4018/978-1-7998-8900-7.ch008

Vaskov, A. G., Lin, Z. Y., Tyagunov, M. G., Shestopalova, T. A., & Deryugina, G. V. (2018). Design of Renewable Sources GIS for ASEAN Countries. In V. Kharchenko & P. Vasant (Eds.), *Handbook of Research on Renewable Energy and Electric Resources for Sustainable Rural Development* (pp. 1–25). Hershey, PA: IGI Global. doi:10.4018/978-1-5225-3867-7.ch001

Vázquez, D. G., & Gil, M. T. (2017). Sustainability in Smart Cities: The Case of Vitoria-Gasteiz (Spain) – A Commitment to a New Urban Paradigm. In L. Carvalho (Ed.), *Handbook of Research on Entrepreneurial Development and Innovation Within Smart Cities* (pp. 248–268). Hershey, PA: IGI Global. doi:10.4018/978-1-5225-1978-2.ch012

Wahab, I. N., & Soonthodu, S. (2018). Geographical Information System in Eco-Tourism. In S. Chaudhuri & N. Ray (Eds.), *GIS Applications in the Tourism and Hospitality Industry* (pp. 61–75). Hershey, PA: IGI Global. doi:10.4018/978-1-5225-5088-4.ch003

Weiss-Randall, D. (2018). Cultivating Environmental Justice. In *Utilizing Innovative Technologies to Address the Public Health Impact of Climate Change: Emerging Research and Opportunities* (pp. 110–143). Hershey, PA: IGI Global. doi:10.4018/978-1-5225-3414-3.ch004

Weiss-Randall, D. (2018). Cultivating Resilience. In *Utilizing Innovative Technologies to Address the Public Health Impact of Climate Change: Emerging Research and Opportunities* (pp. 204–235). Hershey, PA: IGI Global. doi:10.4018/978-1-5225-3414-3.ch007

Weiss-Randall, D. (2018). Climate Change Solutions: Where Do We Go From Here? In *Utilizing Innovative Technologies to Address the Public Health Impact of Climate Change: Emerging Research and Opportunities* (pp. 236–268). Hershey, PA: IGI Global. doi:10.4018/978-1-5225-3414-3.ch008

Whyte, K. P., List, M., Stone, J. V., Grooms, D., Gasteyer, S., Thompson, P. B., . . . Bouri, H. (2018). Uberveillance, Standards, and Anticipation: A Case Study on Nanobiosensors in U.S. Cattle. In I. Management Association (Ed.), Biomedical Engineering: Concepts, Methodologies, Tools, and Applications (pp. 577-596). Hershey, PA: IGI Global. https://doi.org/ doi:10.4018/978-1-5225-3158-6.ch025

Wulff, E. (2017). Data and Operational Oceanography: A Review in Support of Responsible Fisheries and Aquaculture. In P. Diviacco, A. Leadbetter, & H. Glaves (Eds.), *Oceanographic and Marine Cross-Domain Data Management for Sustainable Development* (pp. 303–324). Hershey, PA: IGI Global. doi:10.4018/978-1-5225-0700-0.ch013

Yener, D. (2017). Geographic Information Systems and Its Applications in Marketing Literature. In S. Faiz & K. Mahmoudi (Eds.), *Handbook of Research on Geographic Information Systems Applications and Advancements* (pp. 158–172). Hershey, PA: IGI Global. doi:10.4018/978-1-5225-0937-0.ch006

Yu, K., Liu, Y., & Sharma, A. (2021). Analyze the Effectiveness of the Algorithm for Agricultural Product Delivery Vehicle Routing Problem Based on Mathematical Model. *International Journal of Agricultural and Environmental Information Systems, 12*(3), 26–38. https://doi.org/10.4018/IJAEIS.2021070103

Related References

Zhou, X., Sharma, A., & Mohindru, V. (2021). Research on Linear Programming Algorithm for Mathematical Model of Agricultural Machinery Allocation. *International Journal of Agricultural and Environmental Information Systems*, *12*(3), 1–12. https://doi.org/10.4018/IJAEIS.2021070101

About the Contributors

Abdelkrim Ben Salem, Ph.D., is an Assistant Professor of Ecology and Environment in the Mohammed V University, Faculty of Sciences Rabat, Department of Biology. He joined the University of Mohammed V after ten years as a teacher of qualifying secondary education in Draa Tafilalet Academy. Broadly, her methodological research focuses on vulnerability and adaptation to climate change and valuation of ecosystem services. Within, Ben Salem is currently working on socioeconomic vulnerability of the agriculture and water sector to climate change. He has published interdisciplinary projects across varied outlets, including, Parasites & Vectors Journal, International Journal of Water Resources and Arid Environments and International Publisher of Information Science and Technology Research (IGI-Global).

Laila Rhazi is a professor at the Mohammed V University in Rabat, Morocco. Her research has focused on plant communities dynamics in Mediterranean wetlands in the context of global change as well as on the population biology of threatened species in the ecosystems. Her expertise was built on numerous projects throughout the Mediterranean.

Ahmed Karmaoui started his academic career as Research Fellow at Cadi Ayyad University and awarded Ph.D. from the Faculty of Semlalia (UCA); graduated from Publons Academy, and awarded Top 1% of international Reviewers (2018) by Publons. He has published various peer-reviewed international journal articles, book chapters, and books on a range of environmental topics. He is currently the president of the Southern Center for Culture and Science, Morocco, and an assistant professor at FSTE (UMI). Broad areas of his research interests include ecosystem services assessment.

* * *

Dilara Arslan earned her doctoral degree in community ecology at Tour du Valat Research Institute and l'Université d'Avignon in France. Her doctoral studies focus on conducting a comparative study of biodiversity in the Gediz Delta, İzmir, Turkey. Within her doctoral thesis, she made observations on birds, reptiles and dragonflies. Besides her research activities, she is interested in the locals' cultures and ethnic practices, in addition to photography. She has been a board member of the Mediterranean Conservation Association since 2018 and a project leader of Conservation Activities of Marmaris Salamander.

Issam Boukhris is a Ph.D. candidate in Forest Ecology and Environmental Technologies. His research interests are the use of IoT and Remote sensing for better monitoring of ecosystem evolution. He is currently working on growth modeling of mixed forests using Process-based models.

Asma El Alami is Administrator at the Ministry of Spatial Planning, Urbanism, Housing and City Policy of Morocco. She is also part-time Professor at the Faculty of Sciences Meknès and at Technician Training Institute of Meknès. She accumulated more than 13 years at the public administration to work on several environnement studies like: waste water, solid waste and environmental impact assessments studies. Actually, she continued her researchers on groundwater fauna and their preservation strategies.

Mohammed El Madihi University professor of biology affiliated with the Laboratory of Plant Biotechnology, Ecology and Ecosystem Valorization, URL-CNRST N°10, Faculty of Sciences, University Chouaib Doukkali, El Jadida, Morocco. His research is focused on plant communities in the Mediterranean temporary wetlands in the context of global change and on the population biology of rare and threatened species in this ecosystem.

Saber Er-Riyahi currently works at the Department of Geography, Laboratory of Studies and Research in Geography, Université Moulay Ismail. Er Riyahi does research in Biogeography and Geoinformatics (GIS). Their most recent publication is 'Plant community patterns in Moroccan temporary ponds along latitudinal and anthropogenic disturbance gradients'.

Lisa Ernoul coordinates the Management and Restoration of Natural and Agricultural ecosystems department at the Tour du Valat Foundation, a research centre for the conservation of Mediterranean wetlands. She is involved in several international conservation projects and develops research on human aspects of environmental management around the Mediterranean basin. More specifically, she is interested in the values and perceptions of different populations as they relate to management, decision making and conservation.

Elhoucine Essefi obtained his PhD in geology from the National Engineering School of Sfax in 2013 titled 'Wet Aeolian Sedimentology and Sequence Stratigraphy within the Terrestrial Analogues in Eastern Tunisia: Implications for Wet Aeolian Sedimentology and Sequence Stratigraphy on Mars'. In 2009, obtained the Master degree from the Faculty of Sciences of Sfax titled 'Multidisciplinary study of Sidi El Hani Saline Environment: the Geological History and the Climatic Variability'. In 2003, he obtained the 'Maitrise' degree in Earth Sciences from the Faculty of Sciences of Sfax. In 2011, he obtained the License in Linguistics and Anglophone Civilization from the Faculty of Arts and Humanities of Sfax. In 2015, he obtained the Master's degree in Specialized English titled 'Suggested solutions to some problems of Tunisian novice researchers with academic writing in English: case studies of a Research Article (Gallela et al., 2009) and a Master Thesis (Essefi, 2009)'.

Anis Guelmami is Project Leader/GIS, Remote Sensing and Mapping, Tour du Valat, Research Institute for the Conservation of Mediterranean Wetlands, Le Sambuc – 13 200, France.

M. D. K. L. Gunathilaka is a young academician and researcher with several years of experience in teaching and assisting courses on an undergraduate level. Reviewed several research papers for international peer-reviewed journals on request. Published over 30 abstracts in international and national level conferences and more than 10 articles in peer-reviewed international journals and 02 book chapters and 02 books which are translated into 08 foreign languages. Her PhD research leans towards the aquatic ecology where freshwater lakes are much highlighted.

Soumaya Hajji is an Engineer in Geology from National Engineering School in Sfax (ENIS), Tunisia. She obtained her PhD in Modeling in Hydrogeology in 2015. She obtained her HDR in Environnemntal Hydrogeology and geo-Informatic in 2021. She is a researcher at geochemistry and hydrogeology, modelling and geostatistics at 'Water-Energy-Environment' Research Laboratory (LR3E) in Engineering School Sfax, Tunisia. Her research activities deal with geochemistry, modeling, analysis, optimisation, simulation of groundwater resources, groundwater quality and. She is an Associate Professor in Geology Department in Faculty of Sciences of Sfax, Tunisia.

Younes Hamed is Prof. at the Geological Sciences (Hydrogeology, radioactive isotope geochemistry and Climate change), Univ.of Gafsa Tunisia. He has an academic background in Groundwater Modelling, Environmental Geochemistry & Pollution combined with a practical work experience in geological sciences. His research interests include subsurface exploration, Environmental Contaminants, and Envi. Pollution & risks. He is interested in Reservoirs Production and Estimator quantity.

W. T. S. Harshana graduated from the University of Colombo, Sri Lanka and completed a Diploma in GIS as well. Mr Harshana is currently working on his Master's Degree. Currently, work at the Urban Development Authority where Mr Harshana contributes to various field works and mapping projects.

Merve Kaplan was born in 1986. She was graduated her master's degree from Kocaeli University in Turkey. The subject of her thesis was "Effects of shucking and cooking of striped Venus clam (Chamelea gallina L.,1753) collected from kandira beach with different thermal methods on meat yield chemical composition". She is an aquaculture engineer and she has been working in Kocaeli Provincial Directorate of Agriculture and Forestry, Turkey. Also she is a PhD student at Canakkale Onsekiz Mart University in Turkey.

Hicham Mharzi-Alaoui (PhD and Forest Engineer) is the head of the risk analysis unit at the National Center for Climatic and Forest Risk Management in Rabat, Morocco, He received his Ph.D. degree in Geodetic sciences and engineering, geoinformatics, planning and water from the Hassan II veterinary and agronomic institut in 2017. At the same year, he joined the National school for forest engineer as a Part-time professor. Dr. Mharzi-Alaoui's research focuses on risk analysis, climate change, artificial intelligence, remote sensing and GIS analysis of ecological and environmental systems, land-use and land-cover change, environmental modeling and human-environment interactions.

Said Moukrim has earned his PhD degree in Biological Sciences (Mohammed V University, Morocco) and he is an Agro-economist engineer from the National School of Agriculture of Meknes. Actually, he is an assistant professor of biology at the Faculty of Sciences, Mohammed V University in Rabat. He participated in several projects on the use of new technologies in natural resource management in Morocco. He has published articles and book chapters and has participated in several congress and scientific events in national and international level.

Mouhssine Rhazi has earned his PhD degree in Plant Ecology about the restoration ecology of plant diversity and a rares species in the mediterranean temporary ponds (S. France) at Paul Cézanne University of Marseille- France. Actually, he is a Professor of biology at the Faculty of Sciences and Techniques of Errachidia, Moulay Ismail University of Meknes. He is director of the research team: Biodiversity, Environment & Plant Protection. He participated in several projects focused on the study of the vegetation of the Mediterranean temporary ponds. He has published articles and book chapters and has participated in several congress and scientific events in national and international level.

Seda Yıldırım currently works at Tekirdag Namık Kemal University in Turkey. She studies on sustainable development, 2030 sustainable development goals, sustainable consumption patterns, Islamic finance and banking, consumption patterns, environmental quality and food security. She is married and has one child.

Index

L

M

R

S

T

W

www.ingramcontent.com/pod-product-compliance
Ingram Content Group UK Ltd.
Pitfield, Milton Keynes, MK11 3LW, UK
UKHW050945290926
15470UKWH00005B/1649